Penguin Books
Ground for Concern

Mary Elliott is a legal historian and research assistant at the University of Sydney's Faculty of Law. She has been actively involved in the anti-nuclear movement since the 1950s while a history student at Cambridge University, later to become a founder member of the Uranium Moratorium in Australia. She has written widely: academic papers, historical articles, and many contributions to the media on nuclear issues, including submissions for the Friends of the Earth to the Ranger Uranium Environmental Inquiry.

Friends of the Earth, founded in 1969 in San Francisco, now has groups in thirteen countries including Australia. Its members, who come from many different sections of society, contribute freely their various skills and expertise, bound by common purpose – to protect the environment. FOE Australia has campaigned actively on many environmental issues but because of the magnitude and urgency of the issue, uranium has become FOE's prime concern.

Ground for Concern
Australia's Uranium and Human Survival

Edited by Mary Elliott

Penguin Books

Penguin Books Ltd,
Harmondsworth, Middlesex, England
Penguin Books,
625 Madison Avenue, New York, N.Y. 10022, U.S.A.
Penguin Books Australia Ltd,
Ringwood, Victoria, Australia
Penguin Books Canada Ltd,
2801 John Street, Markham, Ontario, Canada
Penguin Books (N.Z.) Ltd,
182-190 Wairau Road, Auckland 10, New Zealand

First published 1977

This collection copyright © Friends of the Earth, 1977
Copyright in individual contributions is retained by the authors

Made and printed in Australia at
The Dominion Press, Blackburn, Victoria
Set in Plantin by The Dova Type Shop, Melbourne

Except in the United States of America,
this book is sold subject to the condition that
it shall not, by way of trade or otherwise, be lent,
re-sold, hired out, or otherwise circulated without
the publisher's prior consent in any form of
binding or cover other than that in which it is
published and without a similar condition
including this condition being imposed on the
subsequent purchaser

CIP

Ground for concern.

Index.
Bibliography.
ISBN 0 14 004462 0.

1. Uranium mines and mining – Australia –
Addresses, essays, lectures. I. Elliott, Mary,
ed. II. Friends of the Earth.

622.34'93'0994

Contents

List of figures and tables

Foreword
Paul Ehrlich

Preface

Introduction

1 The Nuclear Hazard 7
   Rob Robotham

2 Uranium Mining in Australia 25
   Wieslaw Lichacz   Stephen Myers

3 Reactor Safety 65
   Rob Robotham

4 Nuclear Wastes 97
   Sandy Pulsford

5 Bombs: Nuclear Proliferation and Nuclear Theft 127
   Greg Woods   Mary Elliott

6 A Nuclear-Free World 159
   Jeffrey Nicholls   Michael Bell

7 The Politics of the Nuclear Industry 187
   Herb Fenn

   Conclusion: Stopping the Atomic Juggernaut 209
   Mary Elliott

   Afterword 217
   Judith Wright

   Appendix A 219
   Principal Findings and Recommendations of the First
   Report, October 1976, of the Ranger Uranium
   Environmental Inquiry

   Appendix B 222
   Text of the Nuclear Non-Proliferation Treaty, 1968

References 229
Further Reading 231
Index 233

List of Figures

2.1 Occurrences of uranium in Australia 28
2.2 The diminished Alligator Rivers (Kakadu) National Park, Northern Territory 38
2.3 Alligator Rivers Park proposed by environmentalists 39
2.4 Proposed Ranger mine and treatment plant 49
2.5 Aboriginal interests and land claims 51
3.6 The nuclear fuel cycle 66
3.7 Magnox reactor 71
3.8 High-temperature gas-cooled reactor (HTGR) 74
3.9 Pressurized water reactor (PWR) 75
3.10 Boiling water reactor (BWR) 76
3.11 Cross-section of a BWR 77
3.12 Inside the drywell of a BWR 79
3.13 Fast breeder reactor 80
5.14 Effect of a fifty-megatonne bomb on Sydney, compared with Hiroshima-size bomb 128
5.15 Present and potential nuclear weapons powers 137

List of Tables

2.1 Australian uranium oxide reserves 27
3.2 Nuclear power reactor family tree 72
5.3 States – members of the United Nations – not party to NPT, as at 31 August 1976 140

'Like you I am against the mining of uranium. With Australia becoming increasingly scarred morally as well as physically, I'm surprised there aren't more who share our views. Perhaps there are if they stop to think, and you and your allies will start them doing so.'

Patrick White
From a letter to Mary Elliott, 8 May 1975

Foreword

The people of the world now face a momentous decision: should they proceed to deploy an extremely dangerous nuclear power technology in order to permit nations that already waste huge amounts of energy to waste even more? Australians have a unique opportunity to influence that decision. Although in the long run Australia's uranium could prove to be just a drop in the nuclear bucket, the world is watching to see whether that drop will now be contributed. While the uranium itself would make little difference in the energy economy envisioned by the nuclear establishment, Australia's decision on whether or not to mine and export it could be a key factor in the global choice about to be made.

A decision so important clearly must be made by the Australian people themselves, not by a small cadre of 'experts', be they nuclear scientists and engineers or ecologists. The Ranger Report recognized this in calling for a wide public debate and, fortunately, all of the central issues are easily understandable by lay persons willing to acquaint themselves with the facts.

In my recent visit to Australia, I was dismayed to discover that most of the arguments presented by some of the leading proponents of nuclear power there bore no relationship whatsoever to the facts, but consisted mainly of outright lies mixed with a smear campaign impugning the motives and competence of people who oppose them. The anti-nuclear movement can and should avoid such tactics. The facts of the issue speak for themselves, and people who substitute falsehood and venom for rational discourse do not help their case.

In my opinion the moral decision for Australians would be to leave their uranium in the ground. I think that the risks attached to its removal – such as those of increasing the chances of thermonuclear war and of destroying part of Australia's great natural heritage – are in no way balanced by the purported benefits. But that is one man's opinion. By reading this book, the Ranger Report and arguments raised by nuclear proponents, you can form your own opinion and play a role in what may be the most crucial decision ever made by our species.

Paul Ehrlich
Sydney, 1977

Preface

This book is neither a political manifesto nor a textbook on nuclear power. It is a reasoned statement of the concern that Australians, and people throughout the world, feel about the prospect of a nuclear future. The authors have tried to grapple honestly with the problems of the atomic age, which is our age. They have tried to speak about complex matters in plain language. While the subject does not lend itself to simplicity, it is the premise of this book that people should not be deceived into believing that they cannot understand the issues. These issues affect our very existence and they must not merely be left in the hands of the 'experts'. As the world plunges deeper into authoritarianism, this is a book intended to help people understand the implications for themselves of a nuclear world.

Because it is not an academic treatise, the book is not heavily referenced; rather an attempt has been made to make clear in the text the sources on which it is based. The main exception is chapter four on nuclear wastes, where references to a number of crucial articles and reports are provided because the subject has been so distorted by nuclear proponents that the Australian public may be unaware of the concern that has been expressed in the publications of overseas organizations. A short list of further reading is also provided – books that themselves offer extensive bibliographies of material for those who wish to go into more technical detail on the subject. The contributors to this book, with their disparate backgrounds and qualifications, reflect the wide variety of people who have expressed their grounds for concern about nuclear power to the Fox Commission in particular and within the uranium debate at large. Despite the pressures of many other commitments, the contributors have given themselves to this co-operative venture gladly, and I thank them all for their response.

First to Rob Robotham, now Radiation Protection Officer at the University of Melbourne, who has given the benefit of his experience at Harwell and the Australian Atomic Energy Commission to the writing of the chapters on health hazards in the nuclear industry and reactor safety, many thanks for being our technical anchor-man.

The chapter on the effect of mining on the Northern Territory and on the Aboriginal people was written by two young men who have made intensive study over the last three years of this aspect of the uranium issue. Both are full-time workers for FOE and both have fought the uranium campaign from the beginning. Stephen Myers is currently a stalwart of the FOE office in Darwin. Wieslaw Lichacz won for FOE the status of a principal party at the Fox Inquiry: his knowledge of the evidence presented to the Commission is therefore intimate, and his experience of the Northern Territory and its Aboriginal people is personal.

Another contributor to the Fox Inquiry was Sandy Pulsford, an FOE research officer in Adelaide. The submissions he presented on nuclear waste management provided the basis for his chapter in this book.

The chapter on nuclear weapon proliferation and nuclear terrorism was written by Greg Woods, Senior Lecturer in Criminal Law at Sydney University Law School and myself. It was for Greg Woods that I researched the US Congressional Record for material on nuclear theft, for his paper on nuclear terrorism delivered at the ANZAAS Conference in Hobart in 1975.

Jeff Nicholls presented submissions to the Fox Inquiry on solar energy and is now working on a book covering the social, political and technological implications of solar energy. He was assisted in his contribution by Mike Bell, who having trained in surveying and town planning, has a continuing professional interest in organization of the human environment. In association with Rod Simpson, he also provided material for the illustrations for the book.

The final chapter on the politics of the nuclear industry was written by Herb Fenn, a graduate in political science and worker for FOE in the USA and in Australia for several years.

Those who have put pen to paper have received substantial assistance from many quarters. We are indebted especially to Dale Bridenbaugh, Walt Patterson, Donald McPhee, Ted Wheelwright and Paul Ehrlich. Dale Bridenbaugh, one of the famous 'GE Three' who resigned from General Electric in protest at the dangers of nuclear reactors, gave the benefit of his advice on reactor safety. Walt Patterson's book *Nuclear Power* has done so much to educate us and the public about the nature and danger of nuclear power. He also gave detailed advice on the question of nuclear waste disposal and, like Dale Bridenbaugh, has given us much encouragement. Donald McPhee, geneticist at Latrobe University, has contributed essential advice on the difficult subject of the genetic effects of radiation. Economist Ted Wheelwright at Sydney University also gave his encouragement and advice, especially to Herb Fenn on the operations of the nuclear industry. Paul Ehrlich, who visited Australia in 1976 on a research project, found himself unexpectedly in the middle of the uranium debate. We thank him for the contribution of his fighting words in the foreword to our book.

This combination of scientific knowledge and a sense of social responsibility is a powerful one, but would be incomplete without a contribution from the arts. Two great Australian writers, Patrick White and Judith Wright, have both contributed committed statements to open and close the book.

Wendy Varney has typed the whole manuscript in circumstances which would have defeated a lesser person and contributed her acutely intelligent criticism. My husband, Barry Elliott, has lived it all.

For permission to reproduce material from the *Ranger Environmental Inquiry First Report* and *Second Report*, we thank the Australian Government Publishing Service.

It is a remarkable experience to work with Friends of the Earth. All the people who have helped make this book believe that uranium is ground for concern. We attempt to show why. For doing so, we have been described in a number of sometimes amusing terms – econuts, emotional idealists, communist dupes. This book will give everyone the opportunity to judge the merit and motivation of our cause. To the people of Australia, and to all the people of our planet, Friends of the Earth presents its case.

Mary Elliott
for Friends of the Earth
Sydney, August 1977

Introduction

A few days before its demise in 1972, the Liberal government signed a number of contracts to supply uranium to Japan, West Germany and the USA. That was the beginning of the attempt to mine Australia's large uranium deposits, amounting to 25 per cent of the western world's reserves. The newly elected Labor government, especially its Minister for Minerals and Energy, Rex Connor, was no less anxious to export uranium, but wished to bring the industry under Commonwealth control. This policy, to the chagrin of the mining companies and their multinational parent companies, held up development. Even more distressing to the companies was the passing of the Environment Protection (Impact of Proposals) Act in 1974 which required that, before any major project went ahead, a public inquiry into its environmental effects must be held. This was an historic piece of legislation. It meant that companies came, in an unprecedented way, under public scrutiny. Members of the public, under the procedures of the Act could, in their submissions to environmental inquiries, call to account government departments, company executives, bureaucrats of all kinds.

Under the Act, in July 1975, an inquiry was set up to look into the consequences of mining uranium from Northern Territory deposits held by the Ranger Company. The Commissioners, Mr Justice Fox, Mr Graeme Kelleher and Professor Charles Kerr opened their hearings in the gloomy halls of the old Gaslight Building in September 1975.

They recognized that a full interpretation of the Environment Act required them to consider the effects of uranium mining and export, not only on the immediate environment of the mine, but also on the world at large. To the dismay of the uranium mining companies and the Australian Atomic Energy Commission (AAEC), a partner in the Ranger project, the Fox Inquiry became a thorough examination of the nuclear power question. Its procedure allowed open access to individuals and concerned organizations on a parity with the AAEC and the companies, all witnesses having right of cross-examination. The Inquiry sat for 121 days in Australian capital cities and on the Ranger site. The transcript of the evidence of 303 witnesses amounted to 13 525 pages.

While the Inquiry was sitting the Labor government was dismissed by the Governor-General on 11 November 1975 and the Liberals came back to power. The relationship of the Inquiry with the new government was not a happy one. Its staff was cut and the government, anxious to proceed with mining, ordered the Inquiry to speed up its proceedings. This the Commissioners adamantly refused to do. Moreover, in their First Report, published in October 1976, they complained of resistance by government departments to provision of information for the Inquiry, commenting drily, 'It seemed to us that the objectives and working of the Environment Protection (Impact of Proposals) Act may not be clearly understood in some government departments.' (p. 6)

The First Report covered the general issues of nuclear power safety, nuclear weapon proliferation and the nature of a nuclear society. The Commissioners, under pressure to make public their views, had decided to publish their findings on these issues in a preliminary report before completing a Second Report on the complex question of the effect of mining on the Northern Territory. The findings and recommendations of the First Report are printed as an appendix to this book for all to read since they were grossly misrepresented by both the press and the government. Comment on the content of the Report is made in the text.

The most important recommendation of the First Report was that there should be ample time for public and parliamentary consideration of the Report and discussion of it before any decision was made. The publication date of the Report was 28 October 1976. A mere fortnight later, on 11 November, the Minister for Environment, Mr Kevin Newman, announced that the Report was basically a go-ahead for uranium mining and that existing contracts could now be fulfilled. There was a mere two hours of parliamentary debate on this statement in an almost empty house. There had, obviously, been no public discussion of the Report.

Even before the appearance of the First Report, the government had given permission for export from the Mary Kathleen mine to commence. Many regarded both that and the statement of 11 November as pre-emption of the Inquiry because existing contracts could not be honoured without opening a new mine. Government stockpiles and Mary Kathleen could not provide the 11 000 tonnes contracted: there was a shortfall of about 3000 tonnes. It seemed obvious, therefore, that from the beginning the government, hand in glove with the uranium companies, was determined to export uranium. Though denying in statement after statement that a decision had been made, in practice the Cabinet did little to disguise its aim. Delegations of Japanese, West German, French, Canadian, US and British government officials visited Australia while the Fox Inquiry was sitting. These visits were repaid by the Prime Minister and Minister for National Resources, Mr Anthony, with accompanying Australian officials. Negotiations, taking place at the highest level, were obviously not about the price of eggs.

Meanwhile, however, the public was becoming increasingly aware of the uranium issue. Walt Patterson has described the First Fox Report as a time bomb with a long slow fuse. It presented so much evidence on the hazards of nuclear development, together with the demand that these should be discussed by those they affected, the people, that it was bound to provoke debate. Between the publication of the First Report in October 1976 and that of the Second in May 1977, uranium became a topic of national concern. Unable to avoid public scrutiny, the uranium companies, through their collective organization, the Uranium Producers' Forum, hired a public relations firm to present their case. They spent hundreds of thousands of dollars on television advertising and glossy booklets to be distributed to school children. Together with the AAEC, they sent speakers to meetings being organized by political, religious and community groups. These were but the preliminary skirmishes of the public discussion demanded by the Fox Inquiry. The Commissioners commented at this stage that the level of debate was totally inadequate, and was being conducted as if their Inquiry and the evidence it had collected did not exist. The government and the mining companies now claim that the debate demanded by the Fox Inquiry has in fact taken place. Immediately after the publication of the Second Report, a four-hundred page document, in May 1977, the Prime Minister announced that public debate was at an end. On 25 August, barely three months later, Prime Minister Malcolm Fraser announced in Parliament that uranium mining for export could proceed forthwith. In giving encouragement to other mining companies he rejected the recommendation by the Fox Commission that, if mining were planned to go ahead, the mines should be developed gradually and sequentially.

However, at the Labor Party Conference in Perth in July 1977 a decision of principle was made that will ensure the uranium debate goes on. Almost unanimously the Labor party demanded an indefinite ban on the mining and export of uranium until satisfactory solutions are found for the problems of nuclear waste disposal and the proliferation of nuclear weapons. Most importantly, the Labor conference decided that until such solutions are found it would not be bound, if returned to government, by any new contracts signed by the present government; nor would it open a new mine to honour existing contracts. The Labor stand meant that uranium became a major election issue.

It was the plea of the Fox Inquiry that discussion should take place on a rational, factual basis. The two Reports have been criticized from all sides for not giving a clear line, a green or a red light. But to criticize the Commission thus is to misunderstand its terms of reference, the nature of its main recommendations and the format of the Reports.

The three Commissioners refused to take a decision which they thought should be made by the people of Australia. The Reports comprise a collection of evidence and views on nuclear power and uranium mining. They were written to be read by the general public to assist each person in reaching a decision. They propose certain courses of action that could be taken, outlining dangers and problems and suggesting ways of mitigating those dangers and problems. What they reveal, as this book will demonstrate in detail, is that solutions to the manifold problems of nuclear power are far off, if ever realizable. This is shown clearly in the Second Report which is mainly concerned with ecological problems of mining in the Northern Territory and the effect mining would have on the Aboriginal people. In ten pages of detailed recommendations at the end of the Report, the Commission has proposed so many qualifying measures that must be taken before mining proceeds that it is doubtful whether any commercial enterprise would be possible under these conditions. The Commission rejected the Ranger Company's proposal for its *modus operandi*. It proposed the establishment of independent bodies to control company operations and a research body of eminent scientists to study the region further before any soil is turned. It described in vivid detail both the ecological value of the region and how detrimental the impact of mining would be on the Aborigines.

This book is no substitute for the Reports themselves, but is closely related to the work of the Inquiry and owes a great debt to it. As described in the Preface, the authors have laid before the Inquiry material now presented here for the public, as an introduction to the Reports themselves.

Unlike them, however, this book expresses a definite point of view. It considers the hazards of nuclear power totally unacceptable. The first chapter describes the long-term effects of radioactive materials on health, the genetic effects, the dangers to workers in the nuclear industry, to miners, reactor operators, workers in reprocessing plants. The book therefore opens with the fundamental question of the nature of fission products produced in reactors from uranium and their effect on life.

It moves on in the second chapter to present the Aboriginal case against mining in the Northern Territory and a description of the mining operation with its savage destruction of the local environment. This is the first stage of the nuclear industry; from the beginning uranium is a destroyer.

The third chapter follows the uranium to its destination, the reactor, and considers the much-disputed question of reactor safety. It is followed by a detailed description in chapter four of the many kinds of radioactive wastes that reactors produce, the disposal methods currently under consideration and their inadequacy. This current information will clarify the waste problem that has been confused by conflicting accounts of the present situation.

One of the products of reprocessed wastes is plutonium, the material from which nuclear weapons are made. The subject of weapon proliferation and nuclear terrorism is not one that anyone can contemplate with equanimity, but in chapter five the prospect is faced squarely. It would be dangerous in the extreme to ignore the total inadequacy of the safeguards and treaties governing nuclear materials. Their failings are summarized to complete the case against uranium mining and export.

At this point the reader may feel, quite justifiably, a certain sense of desperation, having followed uranium out of the ground, through the perils of the fuel cycle into the wastes and possibly into nuclear weapons. The next chapter, however, describes another choice open to us, away from a nuclear society and into a nuclear-free world. It questions the costs of a nuclear economy and describes how solar energy in its many forms could become the basic and economic alternative energy source for the world, and the kind of free, self-sufficient society that, given the will and effort, could be created from such a base.

The reader may then ask, why, if this alternative is so much superior in every way to the nuclear option, that billions of dollars were poured into nuclear power. In reply the final chapter explains the political and economic development of the nuclear industry, its growth out of the US war machine, its cost in terms of money and human values and the disregard of profit-making companies for considerations of safety. It comments, without fear or favour, on the Australian nuclear men. The history of the industry supports the central theme of this book, that technocrats must not be allowed to assume the power to make vital decisions beyond public scrutiny.

The conclusion looks ahead to the prospects before us at this decisive turning-point in history and suggests what the individual, who may, wrongly, feel powerless, can do to prevent the world plunging into nuclear catastrophe. As the mining companies prepare to bulldoze their way into the Northern Territory, and the government, regardless of all the hazards described in this book, has finally stated in Parliament its policy to mine and export uranium, it may seem that in Australia the game is lost. Thousands of Australians who have marched in protest through the streets of every Australian city do not agree. Despite intensive efforts to develop the uranium industry by companies and successive governments over the last three years, Australian uranium lies still in the ground. The question of nuclear power, here and overseas, will go on. The grounds for concern on uranium and nuclear power remain unchanged. This book asks the reader to consider them well, for Australia's uranium is Australia's responsibility. What we do now will be judged by the future generations whose world will be our legacy.

1 The Nuclear Hazard
Rob Robotham

Few people reading this book can be unaware of the fact that radioactive materials and the ionizing radiations they emit are hazardous to health. We have come to know this the hard way. The uranium miners of nineteenth-century Europe developed lung cancer as a result of radioactivity in the atmosphere of the Erz Gebirge mines. The early radium workers developed bone and skin cancers, and luminous dial painters who licked their brushes developed cancer of the lips and tongue. The early radiologists who pioneered medical uses of X-rays contracted leukaemia through self-exposure to their equipment. Patients in the 1930s and 1940s who were treated with X-rays for ankylosing spondylitis, or with Thorium-X for tuberculosis, developed leukaemia or bone cancers. American uranium miners of the 1940s and onwards developed and are still developing lung cancer as a result of exposure to radium in the mines. Israeli migrants who were treated with X-rays for ringworm developed thyroid cancer. People who happened to be in the vicinity of Hiroshima or Nagasaki when the first nuclear bombs fell and survived the holocaust, later developed leukaemia or some other form of cancer. Cancer developed in the first ten years of a child's life can be related to the fact that the mothers were X-rayed during pregnancy.

The list above is concerned solely with radiation induction of cancer in humans and is by no means exhaustive. It does, however, show quite clearly that from the earliest days of man's attempts to use radioactive materials for some real benefit (as in the case of diagnostic X-rays) or imagined benefit (as in the case of radium injections for quite trivial indications), the possible risks have not been appreciated at all or have been grossly underestimated. Unfortunately, there is evidence that this is still happening and, worse still, that not all risk assessments produced by proponents of the nuclear power industry are well-informed or even, in some cases, completely honest. This is particularly apparent with respect to what are usually termed the 'genetic' effects of radiation and, since these are the very effects that will have the most important long-term consequences for mankind as a whole, it is vitally important that they be considered with the utmost care.

Genetic Effects of Radiation

By the 'genetic' effects of radiation, we usually mean the ability of radiation to alter the genetic material of living organisms, including human beings, in such a way that the changed or mutant genes are passed on to the offspring. That ionizing radiation has this ability, known as *mutagenicity*, is beyond dispute and, since the vast majority of mutant genes are deleterious for the organism concerned, increased mutation rates resulting from increased exposure to ionizing radiation represent a real threat to the well-being of future generations.

Further development of the nuclear power industry necessarily involves increased exposure of the human population – and all other species, several of which are vital to the continued well-being of man – to increased levels of ionizing radiation. Not surprisingly, proponents of nuclear power prefer to restrict discussion to the sorts of increased exposure that could be expected to result from what they call the 'normal operation of the nuclear fuel cycle'. Risk assessments of this type are therefore based on the very tenuous assumptions that every operation in the nuclear fuel cycle, from the mining and milling of uranium through processing to the high-temperature reactions in the power plant itself and the final disposal of waste, will proceed perfectly according to plan. This must be guaranteed at every facility ever built or disposal site chosen, in every location throughout the world for an indefinite period into the future. The location, once chosen, must never prove unsuitable or be subject to an 'act of God', such as a major earthquake. It is under such ideal circumstances that the risks associated with nuclear power developments are usually assessed for presentation to the public by the proponents of nuclear power.

Even then, it is usually accepted that there will be casualties resulting from such things as *planned* releases of radioactive materials or increased background levels of radiation in certain localities, and figures like 'a few hundred' extra cancer deaths per year and 'a few hundred' extra cases of genetically determined disease per year in the

USA are often quoted. While these numbers seem quite small they are hardly irrelevant to the people affected. However, it is with accidents resulting from human error or misjudgement of potential risk, from materials failure, leakage, natural disasters, sabotage, theft of radioactive materials for military, non-military or terrorist purposes, and the various other scenarios outlined elsewhere in this book, that risks of a completely different order of magnitude must come into the reckoning.

Any one such event could lead to a large release of radioactive materials and, depending on the exact nature of the event, the resulting exposure of ordinary individuals to radiation could have devastating effects on the individuals themselves and on their offspring. Not only are many by-products of the nuclear fuel cycle extremely dangerous for long periods of time (plutonium-239, minute doses of which can produce lung cancer in dogs, has a half-life of 24 000 years and remains extremely dangerous to all forms of life for very much longer than this) but also radiation-induced mutations arising in any one generation may persist unnoticed in the human gene pool for many hundreds of years, ultimately exerting their effects on individuals long into the future. No risk to future generations of comparable magnitude can reasonably be attributed to other available sources of power, and it is for this reason that many people are deeply concerned about the present trend towards a society even temporarily dependent on nuclear energy.

The term 'rem' has been coined as a measurement of the biological damage caused by a unit of radiation exposure – a 'rad'. At very high doses, the effects of radiation are obvious enough. For example, half the people in any group of individuals exposed to 500 rems or more will die almost immediately. Following 200 rems, most people in a group will show overt signs of radiation sickness and many of them will die within a matter of weeks. At lower doses, the effects are much more difficult to isolate and properly understand.

The problem is that exposure of a group of individuals to a relatively low dose of radiation or radioactive material will almost certainly go completely unnoticed at the time. The ore miners of the Erz Gebirge and the uranium miners of more recent times were completely unaware that they were being exposed to a radioactive gas (radon) in the atmosphere of the mines. They would have had no difficulty in breathing, for example, and indeed would have detected nothing at all that might have acted as a warning to their bodies. The first indication of an effect occurred only ten to twenty years after their initial exposure, when a considerable proportion of them developed lung cancer and others developed other forms of cancer. In fact, the relationship between cause (exposure to radon gas) and effect (primarily excess deaths from lung cancer) was not recognized in the case of the Erz Gebirge workers until forty or fifty years later.

We know that this kind of response is exactly typical of relatively low-dose radiation health effects. Firstly, there is almost invariably a long interval between exposure and effect, varying from a few years for radiation-induced leukaemia to twenty or more for solid tumours (e.g., of the lung or brain) and several generations for certain kinds of genetically determined disease. Secondly, not all individuals in the exposed population will be affected and, thirdly, different individuals may be affected in different ways, either because they develop tumours at different sites or because they acquire different types of heritable damage to their reproductive cells. Thus there is no single health effect of exposure to radiation that can be monitored in an exposed population, although in some circumstances there may be a predominance of a particular effect. For example, inhalation of radon gas often leads to lung cancer simply because the lung is the site most in contact with the radioactivity. A similar argument applies to inhalation of plutonium-237 particles.

One often hears the cliché that man has always lived, and indeed evolved, in the presence of background radiation, the unspoken implication being that this exposure has done him no harm at all and therefore a little more is acceptable. It sounds like common sense but it is far from being a commonsense matter. For example, one cannot decide for oneself whether or not the effects of background radiation on man include mutagenesis (i.e., can produce heritable changes to the genetic material) and carcinogenesis (i.e., can produce cancers) solely on the basis of ordinary experience of living, mainly because of the difficulty in grasping the implications of a long latent period, but also because of the complexities of human genetics. The absence of warning signals does not mean there is no danger, especially since the causes of mutagenesis and carcinogenesis provide no such signals. A medical X-ray does not hurt or produce noticeable effects immediately. Higher doses do not hurt either, but will certainly produce cancer.

Another argument commonly used in the nuclear debate is that even if a few mistakes are made and the human race is exposed to much higher levels of radiation, man will evolve resistance and the problem will be solved. Again, it sounds like common sense; again, it is not. The argument is based upon a misunderstanding of processes of evolution; acting on its premises would effectively sentence the human species to extinction. Evolution occurs through natural selection, and natural selection operates at the level of the species, not the individual. Thus more resistant species could be favoured by natural selection. For example, after the atomic explosion of Hiroshima it was noticed that the best survivors were those persistent insects, cockroaches.

As mentioned previously, it is beyond dispute that radiation is both carcinogenic and mutagenic. For the reasons outlined above, however, it is extremely difficult to trace the cause/effect relationship – the long latent period being a particular problem because people will inevitably be exposed to other agents during this time. Compounding this problem is the fact that cancers and mutant genes, like good journalists, seldom reveal their sources. We simply cannot distinguish a radiation-induced cancer from one caused by some other agent such as cigarette smoking, and a newly induced mutant gene that confers, say, severe mental retardation on its unlucky recipient is likely to be no different in any property from similar genes already present in the population. Unfortunately, this masking of the cause/effect relationship provides a loophole for those who, though admitting that radiation causes cancer and mutations, refuse to accept that a particular cancer or deformed child could be the result of radiation. It is interesting to note that when radiation workers develop cancer their employers will often attribute these cancers to 'natural' or other causes including chemicals, while employers in certain chemical and manufacturing industries tend to invoke 'natural' causes or even radiation!

Thus one of the key points about radiation exposure is that it produces health effects that are the same as those already occurring in the population. It used to be thought that cancers and mutations that occurred in the general population had no actual cause and hence were termed 'spontaneous'. This view became current soon after 1928, when Muller first demonstrated that X-rays could produce mutations in the fruit fly Drosophila and subsequently calculated that background radiation alone could not be responsible for mutation rates normally observed in laboratory strains of this species. Nowadays, however, it is widely accepted among geneticists that most so-called 'spontaneous' mutations and cancers do in fact have an external cause and that, while background radiation is responsible for a proportion of them, other environmental factors (mainly chemicals, both naturally occurring and synthetic) are responsible for much of the remainder. On this view, any extra radiation dose above natural background levels will therefore produce extra cases of cancer and genetically determined disease. The problem thus becomes one of degree. This is why proponents of nuclear power like to restrict discussion to events associated with the 'normal operation of the nuclear fuel cycle' and to present rather than possible future scales.

The frequently reiterated argument that the coal-based power industry is just as dangerous as the nuclear power industry depends for its thrust on taking into account only events occurring in the 'normal' (i.e., perfect) operation of the nuclear fuel cycle, and ignores completely the sorts of considerations outlined thus far above. Equally, the often repeated statement that no one has yet died as a result of a nuclear accident in the (embryonic) nuclear power industry depends for its thrust upon ignoring the delayed effects of radiation on humans (such as cancer and genetic damage), and tacitly assumes that only the immediately lethal effects of relatively high doses of radiation are of any consequence or interest to the public at large. To talk only in terms of 'accidents' in the nuclear industry conveniently excludes the hundreds of uranium miners who have died of lung cancer, since their deaths resulted from misjudgement, not accident. Those in control of the nuclear industry have not until recently looked in any systematic fashion for evidence of any delayed effects of radiation among their employees. Even if they had been doing this all along it is probably still too early to detect such effects as the induction of solid tumours, which may take two to three decades after exposure to become apparent, or recessive genetic damage which will take several generations before being expressed as genetically determined diseases or deformities.

Among the most misleading statements about the long-term effects of radiation have been several made in recent times – almost invariably by non-biologists – about survivors of the Hiroshima and Nagasaki bombs. For example the Melbourne *Herald* of 20 April 1977 reported: 'There is absolutely no scientific evidence of genetic deformities arising from the bombing of Hiroshima or Nagasaki – and the time lapse is now sufficient.' Such statements are always worded very carefully. Note, for example, that the statement does not claim that there is no evidence of delayed health effects, only that there appears to be no evidence of 'genetic deformities' – as if to imply that only visible (and presumably gross) physical abnormalities are of concern as a result of damage to the genetic material.

This is misleading in the extreme, since only a low proportion of genetic damage will ever be expressed as heritable (physical) deformities. On the contrary, the range of possible effects of a mutagenic agent such as ionizing radiation on an organism as complex as man is enormous. Mutations are produced completely at random in the genetic material, and the resulting mutant genes can have a wide variety of effects on the exposed individuals themselves or their offspring. Whether it is the exposed individuals that are affected or their offspring depends solely on the kind of cell in which a mutation is produced. Mutations arising in what are termed somatic cells (the ordinary cells of the body such as skin cells, lung cells, bone marrow cells, and so on) affect only the irradiated individual. By contrast, mutations arising in certain cells of the reproductive organs termed the germ cells (spermatocytes in males and oöcytes in females) are transmitted to, and only expressed in, the offspring. Even then, mutations will only be expressed and transmitted to future generations if the damage to the genetic material is not so severe as to be incompatible with survival of the foetus. This still leaves an enormous range of possible health effects on future generations, many of which are quite subtle and hence not easily recognized.

A further complication is that some mutant genes are dominant, while the others are recessive. In the former case, induced mutations will be expressed in the immediate offspring of irradiated individuals and hence one might expect to be able to detect them relatively soon after the irradiation. In the latter case, expression depends upon inheritance of one copy of the mutant gene in question from one parent and a second copy, also mutant, from the other. This may sound like an unlikely event, but a far-sighted view will show how serious are the implications for the future well-being of mankind.

Firstly, the human gene pool already contains a significant proportion of recessive mutations, so that pairing of two such genes, one or both of which were radiation-induced, is not at all unlikely over a number of generations of random mating. Secondly, recessive mutations once introduced into the genetic material cannot be repaired in any way but rather will tend to persist for hundreds or thousands of years in the gene pool. It is quite absurd to expect to be able to recognize effects depending upon this type of inheritance in the short time that has elapsed since Hiroshima and Nagasaki, and yet it is these very effects of ionizing radiation that are the major concern of geneticists, partly because of their insidious nature and partly because of their persistent capacity for harm. Thus when it is stated, as it is in the quotation given above, that 'the time lapse is now sufficient' it should be borne in mind that the question of persistent recessive mutations – the vitally important one for future generations – is being totally ignored.

In the light of this, it is worth considering what is already known and what one might expect to be able to observe as a result of the Hiroshima and Nagasaki explosions.

It is known, for example, that many of the people who were in the areas concerned at the time received significant doses of ionizing radiation and yet survived. This is not unexpected after doses of less than about 200 rems or so. However, since radiation is known to be mutagenic at dose levels much lower than this, one would expect new mutations to have been produced in a proportion of these individuals. Some of these mutations would have been produced in somatic cells, others in germ cells. The former type should lead to excess cases of cancer appearing among the survivors, since the great majority of human cancers are now generally believed to be a direct result of somatic mutations. The evidence that there has indeed been an increase in cancer incidence among the Hiroshima survivors is *unequivocal*. At first the evidence was restricted to leukaemia (which has a relatively short latent period), but as time went on excess cases of many other types of cancer with longer latent periods (e.g., solid tumours) have been observed. In addition, it has been known since

1966 that excessive amounts of chromosome damage have been incurred by survivors of the bomb. Thus there is incontrovertible evidence of damage being done to somatic cell genes of the population concerned.

The position with respect to germ cell genetic material is much more difficult to assess. As previously mentioned, the mutations produced in such cells can be either dominant or recessive. Dominant mutations can produce such effects as gross physical deformities, dwarfism, blindness, mental retardation, etc., but it should be remembered that essentially any organ, function or structural component of the body can be affected. Because there are many hundreds of possible outcomes – almost all of which are deleterious for the individuals concerned – of new dominant mutations arising in the relatively small population of bomb survivors, it has not been possible to demonstrate conclusively any increase in particular abnormalities in the first generation to be conceived after exposure. Dominant conditions have undoubtedly been observed, but as they do not identify themselves with respect to cause, there is no way of knowing whether or not they resulted from the radiation exposure. Dominant mutations are known to occur much less frequently than recessives, however. Since the latter can also affect every biochemical process, tissue, organ and structure of the body (once again the vast majority are deleterious), but are not necessarily expressed for several generations, it is most unlikely that we shall be able to obtain direct evidence of their production following the Hiroshima and Nagasaki explosions.

None of these difficulties alters the fact that ionizing radiation is known to be mutagenic. The problems inherent in the situation simply provide room for the unscrupulous to mislead others into a false sense of security. Thus, although there is direct evidence that survivors of Hiroshima and Nagasaki incurred damage to their somatic cell genes, the effects on their germ cells and hence their offspring are likely to remain, tragically, more real than apparent.

The seriousness of a steady, unseen accumulation of new recessive mutations in the population can hardly be over-emphasized. There is after all a limit to the proportion of such mutations that can be tolerated in the human species without there being a drastic deterioration in fitness to survive. Though the accumulation process cannot be observed in the present, future generations will certainly feel the effects.

The accumulation process will certainly be accelerated by increased radiation exposure, whether it be exposure to the low levels we are asked to believe will never be exceeded for any reason, or to the high levels that will follow major accidents or nuclear conflict. After all, as pointed out by the International Commission on Radiological Protection in its 1965 report, it is immaterial in the long run whether defective genes are introduced into the population by many individuals who have received small doses of radiation, or by a few individuals who have received correspondingly higher doses. For future generations, the prospect is: heads we lose, tails we lose.

Health Hazards in the Nuclear Industry

Mining

The first evidence of lung cancer in uranium miners appeared in the 1920s and 1930s after medical investigation of miners working deposits in Joachimsthal. Similar effects have appeared in miners working uranium deposits in the USA after the second world war. Inadequate ventilation and insufficient expenditure on safety have been blamed for the lung cancer deaths of over one hundred American uranium miners. A study by the US Public Health Service has estimated that from 600 to 1100 miners will die of lung cancer induced as a result of radon exposure in the mines. Because of the latency period between radiation exposure and subsequent appearance of the disease, the miners have already exceeded the dose required to produce lung cancer and nothing can now be done to prevent it developing.

In Australia most mining to date has been 'open-cut' and it has been assumed that radon concentrations at this type of mine do not build up to hazardous levels. But during previous mining very few measurements were made of radon levels in air, and no follow-up studies have been made on the workers to see if there has been an increased incidence of lung cancer among them. The radon levels may have been within acceptable levels; there may not have been an increased incidence of lung cancer: it is impossible to tell at this stage. The problem of radon exposure is essentially one of control: if the miners' exposure levels are kept below prescribed limits, the likelihood of developing lung cancer is substantially reduced. To cope with the problem, radiation protection standards, based on the relation between the estimated intake of the decay products and the incidence of lung cancer among underground miners, have been adopted by some countries including the USA but not, at the time of writing, by Australia.

The implementation of codes of practice, safety standards and the continuing assessment of exposure levels and regular monitoring of the health of the workers are all essential if the risks in uranium mining are to be kept under control. It is vital, however, that an independent body be given the task of overseeing the overall safety of the operation. This body must be free of political and financial pressure and must be given effective power either to shut down mining operations if safe levels are continually being exceeded or to require modification of plant and working procedures in the event of unsatisfactory safety conditions. In essence, this is what the Second Fox Report has recommended.

During the milling of uranium ore into yellowcake, radon and dust are present. This operation takes place under more controlled conditions than does the mining. The dust can be controlled to some extent by ventilation and water sprays, but it is still important to measure radon levels in air regularly to check that allowable levels are not being exceeded. The radiation exposure of workers can be controlled by limiting the length of time they spend in higher level radon areas.

Yellowcake is only slightly radioactive, as most of the uranium decay products are removed from the ore during the milling process. It presents special handling problems only if it is not treated as carefully as any other toxic chemical. When quantities are stored in bulk, however, the radiation dose rate may require limitations on workers' access to storage areas.

Tailings

Milling converts the original ore and the associated volumes of non-radioactive rock into a finely ground sand called tailings. These contain all the radioactive decay products of the uranium that were responsible for most of the radioactivity in the original ore. There are vast piles of uranium mine tailings scattered around the world.

In the USA the military rush for uranium has bred an estimated 90 million tonnes of tailings, much of it piled on riverbanks in the southwestern USA. The consequent radioactive pollution of waterways represents a serious problem. At one stage people living downstream in the Colorado River basin were exposed to three times the maximum permissible intake of radium recommended by the International Commission on Radiological Protection (ICRP), because of the radioactive contamination of drinking water from the tailings dumps. Radon will be released from the tailings piles long after mining has ceased. What is more, the radon will be dispersed much more readily than it would had mining not taken place, for the milling process reduces the rock to fine sand from which the radon can much more easily escape.

Professor Robert Pohl of Cornell University in the USA has calculated the long-term health effects from this continuing release of radon on mainland USA. Relating his findings to comparative data suggests that the short-term health hazards of nuclear energy production, including deaths from lung cancer caused by tailings radon, are considerably less serious than those hazards to the health of the general public from coal-fuelled electricity, such as atmospheric pollution, industrial accidents and black-lung disease. However, the long-term releases of radon are not taken into account. The comparative figures for coal were derived from data produced by the US Environmental Protection Agency and no similar figures are available for Australia. The health effect in Australia should be even less because of the lower population density of northern Australia, but it is worth noting that, at a relatively modest wind speed of twenty kilometres per hour, the radon could travel about 3000 kilometres during its mean life. Thus, under the appropriate wind directions it is capable of affecting the heavily populated Indonesian islands. It appears that public health is endangered by electricity production both from nuclear power and from coal – with one important difference: with the latter the using generation pays the cost; for nuclear energy we appear to be prepared to let future generations pay. (For the nuclear industry's poor record on the control of tailings, see Chapter 4.)

'Enrichment'

The enrichment process of uranium is the increase in proportion of fissile uranium-235 to non-fissile uranium-238, by separating out and concentrating the former for more effective chain reaction and therefore greater economy. For nuclear weapons it is often necessary to have uranium that is 90 per cent uranium-235. While not directly as hazardous to health as some stages of the nuclear fuel cycle, enrichment also has its dangers.

There are several possible techniques, called 'gas diffusion', which require the conversion of uranium oxides into the gas uranium hexafluoride, ('hex') – 'a viciously corrosive, reactive gas, requiring very careful handling and high-quality metallurgy in the vessels through which it travels'. In one process, separation and concentration of the required isotope is effected by repeated diffusion of hex through porous membranes. The technology entails thousands of pumps, condensers and large-scale cooling systems, making gas diffusion plants among the hugest industrial complexes in existence. Other techniques include the use of successive centrifuges, or of deflection from gas sprayed from nozzles. Because of the military interest in enrichment for weapons applications, much secrecy surrounds these plants and their processes.

The accidental release of uranium hexafluoride is probably the greatest potential hazard of the enrichment process, although a new technique using lasers presently being developed could prove a radically more efficient method of separating out uranium-235, thereby providing a much more direct route to obtaining weapons-grade material, even from uranium ore. No doubt the details will be kept secret. Perhaps it is some small comfort to know that it is only the centrifuge kind of plant that is periodically planned for Australia, the AAEC having been working on the concept for some considerable time. A spinning centrifuge could accidentally release uranium hexafluoride, but the hazard, we are assured, is no greater than the kind experienced by large-scale chemical plants.

Fuel Fabrication

The next stage in the nuclear fuel cycle is fuel fabrication. Most fuel now used in current power reactors consists of ceramic pellets of uranium oxide contained in metal cylinders. Mixed oxide fuel (MOX), in which fissile plutonium oxide is mixed with natural or depleted uranium oxide to give a fuel equivalent to enriched uranium, may become important in the future. The production of uranium dioxide fuel elements is now well established and seems to be free of significant hazards. Manufacture of MOX elements seems to be considerably more fraught with hazards. The toxicity of the plutonium involved makes it essential that work on it be carried out by operators working outside totally sealed cells. In addition the critical mass of plutonium – the quantity in which a chain reaction could occur spontaneously – is only a few kilograms, so great care has to be taken to control the quantities being worked on at any one time.

Reactors

The radiation hazards during fuel production can usually be kept under control by suitable management techniques. During the first stages of the fuel cycle, therefore, the hazards are relatively containable. It is at the point when the fuel is fed into the reactor that more serious problems begin. While a nuclear power reactor is operating, vast quantities of radioactive materials are present in the reactor core. The process of nuclear fission produces a wide range of fission products. The neutron bombardment of the reactor materials, cooling medium, etc., induces radioactivity in those materials. And all the time the uranium-238 molecules are lapping up stray neutrons and producing plutonium-239. The radiation hazards from reactor operation can arise from either routine or accidental releases of radioactive materials during normal operations, or from the massive release of large quantities of radio-nuclides that would follow an accident involving the reactor core.

Wastes

During their normal operation, reactors produce a variety of low-activity wastes (see Chapter 4). The solids are usually buried, the liquids discharged to streams or the ocean, and the gases released to the atmosphere. The major gases involved are iodine-131, iodine-129, argon-41, tritium (H-3) and carbon-14. Although, with the present number of nuclear stations, the routine release of these radioisotopes does not appear to present a significant hazard to the general population, any major expansion of reactor-building must put the population at large at proportionately higher risk of radiation exposure. Already in the USA there is increasing concern about the steady release of iodine-129 which has a half-life of 16 million years.

Thermal pollution from nuclear reactors is also a major environmental hazard. Only about one third of the heat produced in a reactor is converted to electricity. The remainder appears as waste heat and has to be dumped in the local environment. The quantity of heat is very great; if it is released into a river or lake it may raise the water temperature by several degrees Celsius, with marked effects on the ecology of the water.

Low-level solid wastes generated at nuclear power sites amount to quite significant volumes. In Britain for instance the Central Electricity Generating Board (CEGB) expect that about 2400 cubic metres of solid waste will be accumulated at each reactor site during its working life. Half the total volume, containing the bulk of the radioactivity, comprises metallic components that have been removed from the reactor core, such as fuel stringer bits and pieces, control rods, chains and neutron-measuring instruments. They have to be kept in concrete storage vaults in the reactor as they are intensely radioactive.

Reprocessing

But the greatest problems are still to be encountered in this odyssey through the fuel cycle. We now get to the sowing of the dragon's teeth: fuel reprocessing and waste management. Uranium cannot be left in the reactor until it is all 'burnt'. In practice, only about one third of the fuel can be used before the fuel element has to be removed. It is taken to a reprocessing plant where the cladding is stripped off and the spent fuel dissolved in strong nitric acid. The acid stream is then divided into three components – the potentially 'useful' 'unburnt' uranium and plutonium, and those troublesome left-overs, the fission products. The gaseous fission products contained in the fuel elements – notably krypton-85, iodine-131 and iodine-129 – are released from the fuel pellets during reprocessing. Krypton-85 (half-life: 10.8 years) and tritium (half-life: 12.3 years) are vented routinely from existing plants but the output of these two gases will almost certainly have to be removed as more fuel is reprocessed.

At the time of writing, no major oxide fuel reprocessing plant is operating in the western world – because of engineering problems, licensing difficulties and expense. Reprocessing in the USA has a sorry history. Attempts have been made to establish three reprocessing works in that country; only one has ever operated and that one has been closed since 1972 and may never reopen. A main reason has been its inability to stay within the permitted levels of radiation to workers. The plant that carried out reprocessing from 1966 to 1972 is in Buffalo, New York. It employed an average of 1400 'supplemental' or temporary workers each year during its five and a half-year operating life; that is more than ten temporary workers for every permanent one. Some of these, who were as young as eighteen, worked for as little as three minutes, during which their bodies absorbed the maximum amount of radiation allowed in three months. They were given half a day's pay and were sent on their way. Sometimes six men would be used, one after the other, to tighten or remove one bolt, and occasionally fifty to a hundred welders would be used to complete a small number of welds in a radiation environment. In short, workers were not used for their skill, but for their bodies' ability to absorb radiation.

It is therefore obvious that the final stages of the nuclear fuel cycle are fraught with health hazards. How the dangers of reprocessing nuclear fuel can be coped with remains unclear. The ultimate problem, that of safe waste disposal, is of such magnitude as to require a chapter of its own (Chapter 4).

The Plutonium Legacy

The core of the waste problem is the production, through fission in the reactor, of high-level, long-lived substances which in minute quantities can cause fatal cancers. Of these the most infamous is plutonium. About plutonium the First Fox Report says:

As a poison, plutonium in small amounts acts by causing cancer. In soluble form it can be taken into the body through the digestive tract or through a cut, and may produce cancer in the tissue where it is localized. It is perhaps even more potent when it is in the form of tiny insoluble dust particles, minute quantities of which may cause lung cancer. Unlike the effects of potent chemical poisons, which take effect immediately, the cancers caused by plutonium are not likely to develop for many years.

Plutonium resembles radium in its toxicity and mode of action. The ICRP and the UK Medical Research Council recommend less than a millionth of a gram of plutonium-239 as the maximum amount that should be retained in the body of any person working with the substance. (pp. 88-9)

Plutonium separation in a fuel reprocessing plant is at present about 99 per cent. If this is improved by a factor of ten – a significant improvement in chemical engineering technology – separation will then be 99.9 per cent. The other 0.1 per cent will appear as plutonium contamination of the fission product stream. A 1000-megawatt (electrical) reactor produces in one year about 230 kilograms of plutonium. A worldwide programme of 3000 reactors by the year 2000 will be producing 700 000 kilograms each year. So about seventy kilograms (or 0.1 per cent) will be contaminating the fission products. Call this, for convenience, 70×10^9 micrograms. It is anticipated that the world population by the year 2000 will be 7×10^3. So we will be committing to storage an average of ten micrograms for every man, woman and child on the planet. This is ten times the maximum permissible intake recommended by ICRP. This is not to suggest that we will live in a new egalitarian world where every person will get his or her fair share of plutonium, but the calculation does serve to indicate the sorts of quantities with which we shall shortly be dealing. And this is assuming some way is found to deal safely with the other 99.9 per cent.

It is obvious that, as is generally recognized, at all stages of the nuclear fuel cycle both immediate and long-lasting hazards to life are created. Proponents of nuclear power persistently deny that they exist, even maintaining that uranium ore buried in the ground is as dangerous as the fission products produced in reactors and widely dispersed throughout the world. But at Hiroshima there are, preserved in jars, the tiny bodies of horribly deformed babies born after their mothers had been irradiated by the atomic bomb – preserved as a warning to the world of the radiation hazard. That warning has been ignored. By the time full scientific data on the exact number of cancers and mutations produced by nuclear materials could be collected, a feat that would require an unprecedented surveillance system covering several centuries, the human race, genetically, would be beyond repair.

To deny the undoubted toxicity of materials like plutonium, to create them throughout the world as part of the normal, commercial production of power, and without knowing their full, long-term (especially genetic) effect, is the height of irresponsibility. The human organism, made up of complex cells, transmitting through heredity the effects of the environment prevailing on them, will certainly be threatened by contact with radioactive products of nuclear power, should they be let loose. The little we know about plutonium and other radioactive substances is already frightening; what we or our innocent progeny may learn of their effects is terrifying. Yet they are already being produced in quantity on a commercial scale and Australia's uranium will be instrumental in increasing that scale. No one is immune to the effects of radioactivity; once we have created these substances, they are here to stay for a time beyond human experience or imagination.

2 Uranium Mining in Australia
Wieslaw Lichacz
Stephen Myers

Uranium is not a rare element in the earth's crust. It occurs in proportions of two to four parts per million in most rocks. But it is a heavy element which, over 2000 million years of geological time, has formed concentrated deposits in specific areas. Australia, one of the oldest geological shield areas of the planet, possesses rich deposits of uranium. As a heavy 'overburden' of rock debris, sands and silts, built up over these uranium deposits, they were compressed into a hard matrix of uranium-bearing rock. As the land surface was eroded by the action of winds and rivers, the concentrations of radioactive ore were exposed, releasing small quantities of radioactive material into the environment. The natural erosion of a body of ore may occur over a period of a thousand million years, at a rate so slow that most biological systems are not upset by it. But extraction of the ore by man could release its radioactivity into the environment over the period of a single generation. Whether the balance of nature could survive such a shock is ground for concern.

Uranium deposits were first recorded in 1894 at Carcoar in New South Wales. In 1906 the Radium Hill deposit in South Australia and in 1910 the Mount Painter deposit, also in South Australia, were discovered. Until 1934 spasmodic mining took place at these sites for radium, the heavy radioactive metal associated with uranium. But an extensive exploration programme began only as a response to development of atomic weapons by the Allies during the second world war. Although Australian uranium was not used in the Hiroshima and Nagasaki bombs, they did signal the birth of the Australian uranium industry. The feverish programme of exploration that followed resembled a gold rush. The federal government promised tax-free rewards of up to £25 000 for uranium ore discoveries and set up a uranium ore-buying pool. In 1952 the Income Tax Assessment Act was amended to exempt from tax those profits earned from mining and treating uranium by predominantly Australian controlled companies. This exemption was extended subsequently to income received from uranium mining by taxpayers resident in Australia.

Major discoveries followed at Mary Kathleen in north-west Queensland and at Rum Jungle, ninety kilometres south of Darwin, where a prospector/farmer, Mr J. White, stumbled upon the deposit while killing kangaroos. The Rum Jungle deposit was on freehold land and in 1950 the government took control of the land under the powers of the Atomic Energy (Control of Materials) Act of 1946. In August 1952 the Minister for Defence declared the mining and processing of uranium to be a special defence undertaking. The presence of a high-grade orebody at Rum Jungle had been verified by that time and Rum Jungle was declared a prohibited area. It was during this year that a party of experts from the UK and the USA arrived to discuss uranium supplies.

In January 1953 the government signed an agreement with the Combined Development Agency (CDA), a joint US/UK purchasing agency, whereby CDA provided the capital funds for the mine and Australia provided the uranium over a ten-year period. In the same year, Consolidated Zinc Pty Ltd was authorized to operate the mine on behalf of the Commonwealth of Australia. Consolidated Zinc (which merged with Rio Tinto Mining Company of Australia in 1962 to form Conzinc Rio Tinto of Australia) then formed a wholly owned subsidiary, Territory Enterprises Pty Ltd, for the purpose of carrying out the work. In April 1953 the Australian Atomic Energy Commission (AAEC) was established under the provisions of the Atomic Energy Act and immediately assumed control of the project. Exploration activities reached a peak just after the passing of the Atomic Energy Act. The Act vested the ownership of all 'prescribed substances', which included uranium and all its daughter products, in the Australian government through the AAEC. No uranium could be exported without government approval.

But in the 1960s the weapons programmes of the USA and the UK were cut back and planned nuclear power projects were delayed. The world demand for uranium dropped and accumulation of uranium stockpiles commenced. In 1966 a ban was imposed by the USA on imports to protect US producers. The US base price for uranium oxide fell from about US$8 per pound (US$17.60 per kilo) to US$5.50 between 1968 and 1971. By this time, over a period of ten years, approximately 7600 tonnes of uranium oxide had been produced from Mary Kathleen, Rum Jungle, South Alligator River (Northern Territory) and Radium Hill (South Australia). Between 1961 and 1964 all mines and treatment plants, except for Rum Jungle, were closed and exports ceased in 1965 when contracts expired. Production from stockpiled ore continued at Rum Jungle and the uranium oxide was stored by the AAEC. In 1971 the Rum Jungle plant was dismantled and sold.

Therefore, exploration in Australia was at a standstill by the late 1960s. At this time the Mary Kathleen mine alone accounted for 70 per cent of Australia's low-cost proven reserves. But, in a bid to stimulate exploration, the federal government guaranteed approval to the export of specified quantities of uranium oxide, depending on the size of new deposits. By 1971 prolific discoveries had been made and quantitative restrictions on exports were lifted. Australian uranium oxide reserves increased approximately twenty-three-fold between 1967 and 1974 (reserves of uranium oxide extractable at up to US$15 per lb [US$33 per kilo]).

Table 2.1

Australian uranium oxide reserves (tonnes)

	1967	June 1974
Reasonably assured	10 541	221 716
Possible, but not proven	2 839	96 706
Total	13 380	318 422

Source: Bank of New South Wales Review, no. 14, April 1975 (since metricated)

Over 70 per cent of these reserves are located in the Alligator Rivers area of the Northern Territory, in a series of rocks known as the Koolpin Equivalent extending from east of the Mt Partridge Range northeast to Coopers Creek. In 1969 no minerals of commercial value were known in the area. In the early 1970s, following the issue of permission to prospect, major uranium deposits were discovered in five localities. Virtually the whole of the area between the East and South Alligator Rivers, together with a substantial area to the east within the Arnhem Land Reserve, was covered by exploration licences. The four major deposits, Ranger, Jabiluka, Koongarra and Nabarlek, are all on

28 Ground for Concern

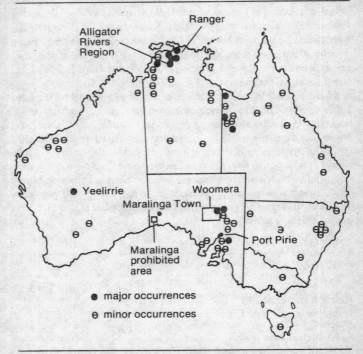

Figure 2.1 Occurrences of uranium in Australia as at 1976

- ● major occurrences
- ⊖ minor occurrences

Source: Compiled by Mike Bell and Rod Simpson from AAEC annual reports and information provided to the Fox Inquiry by the Department of National Resources

lowland country. The Ranger deposit, held by a joint venture of Peko Mines N.L. and Electrolytic Zinc Co. of Australia Ltd, lies at the foot of Mt Brockman. Tributaries of Magela Creek pass through there and the deposit may extend to the alluvial land system just to the north of the Mt Brockman sandstone massif. The Ranger site lies close to two important Aboriginal sacred sites, Djidbi-Djidbi and Dabji. The Jabiluka deposit, about twenty-four kilometres north of Ranger, is held by Pancontinental Mining Ltd. This deposit up till now has the dubious honour of being the world's largest reserve. (Recent discoveries at Key Lake, Canada, may rival it. The announcement of this find in February 1977 sent the 'glamour' stock, Pancontinental, down on the sharemarket.) It is situated on a narrow strip 500 to 800 metres wide between the escarpment and the proposed national park boundary, in an area of paperbark forest. Jabiluka's new deposit may extend under the escarpment and into the proposed park. Aboriginal archaeological and art sites occur in sandstone residuals within a distance of one kilometre of the deposit. The Koongarra deposit, held by Noranda Australia Ltd, occurs on a strip of lowland country about one kilometre wide between a sandstone residual and a tributary of Nourlangie Creek. The deposit situated in the heart of the proposed national park, twenty-four kilometres south-west of Ranger, extends to a rocky foot slope of sandstone. The Aboriginal art sites associated with Nourlangie Rock are four kilometres west of the deposit. The Nabarlek deposit, in the Arnhem Land Aboriginal Reserve, is held by Queensland Mines. The orebody is less than 200 metres from an Aboriginal sacred site known as Harbo Djang, the dreaming place of the green ants.

The largest deposit found outside the Northern Territory is Yeelirrie, eighty kilometres south-west of Wiluna in Western Australia, which is leased by Western Mining Corporation. Other recent finds of uranium concentrations have been made at Ngala Basin, 300 kilometres north-west of Alice Springs, at Pandanus Creek, fifteen kilometres west of the Queensland border in the Northern Territory, and on the north-eastern coast of Queensland near Townsville.

Confusion can be created by use of the jargon associated with uranium exploration. Such terms as 'anomalies', 'resources', 'estimated additional resources', 'reasonably assured resources', 'ore bodies' and 'reserves' can be manipulated to give varying estimates of Australia's actual uranium potential. Politicians are adept at using these little-understood terms in order to show that Australia might receive enormous financial gain from uranium development – estimates that usually disregard the full costs of development, direct and indirect, to the taxpayer. It is therefore important to understand the jargon in order to judge the extent of Australia's uranium 'bonanza'.

At the beginning of the exploration process, the prospector sets out to search for what are termed anomalies. The usual method is by air survey, using magnetic and radiometric techniques to record systematically the changes in magnetic field and radiations emanating from the ground. Sometimes ground surveys using geiger counters are employed as a back-up. In some cases the radiations will not penetrate the overburden and it is necessary to do some test drilling if an area suspected to contain an anomaly is to be proved. One ingenious method of prospecting for uranium is to examine the material brought to the surface by ants while building their large anthills across the sandy plains. The thicker the overburden, the more costly and difficult extraction will be. 'Resources' of uranium can refer to trace quantities of uranium in the soil, rock and oceans, which may not be economic to develop. The term can be used by people supporting uranium exploitation to overestimate financial gains. What matters is whether extraction of these resources is practical and economic. 'Estimated additional resources' refers to uranium that is predicted to exist in unexplored extensions of known deposits, or in undiscovered deposits in known uranium districts. 'Reasonably assured resources' refers to uranium, occurring in known ore deposits, recoverable with currently proven mining and processing technology. Once an anomaly is defined or delineated in size, ease of extraction and grade of ore, then it is classified as an orebody. The term 'reserves' is applied only to reasonably assured resources currently included in the cost category below US$15 per pound (US$33 per kilo) of uranium oxide. Cost categories are based on estimates that include the direct cost of mining, milling and extraction and the writing-off of capital used in providing and maintaining production units. All profits and exploration costs are excluded.

The First Report of the Ranger Uranium Environmental Inquiry estimated that Australia possesses 70 per cent of the western world's uncommitted, economically recoverable, reserves of uranium. How significant the effect of releasing this amount onto the world market would prove is a matter of controversy, as are the economics of development, the cost to the Australian taxpayer and the return he or she can expect on the investment. These crucial questions are dealt with later in the book. What concerns us here is the effect of uranium mining on the people and environment of the Northern Territory, as yet a relatively untouched area of great natural beauty and traditional land of Aboriginal peoples. Would uranium mining, as

Patrick White says, morally and physically scar Australia and in particular the Northern Territory? Our answer is, resoundingly, yes.

The Record of Mining in the Northern Territory

From its inception to the present day, the mining industry in the Territory has had a history of disregard for the environment. This attitude has led to dead, pock-marked landscapes and polluted watercourses. In the pioneering days when little was known about the environmental impact of mining, little could be expected in terms of environmental protection. But with the growth in knowledge about the delicacy of the balance of nature, and humankind's dependence on the maintenance of that balance, a change of attitude might have been expected. Yet, over the last two decades, and right up to the present, major environmental atrocities have been committed in the Northern Territory nowhere more so than at Rum Jungle. There the miners have been and gone, leaving an example of devastation that should provide an object lesson on the immediate environmental effect of uranium mining. As a result of the Rum Jungle experience, we can no longer plead ignorance of the laws of nature which such an enterprise so crudely flaunts.

At Rum Jungle three separate open-cut mines were worked in the area, as part of the overall operation. White's Cut and Dyson's Cut were completed in 1958 and the third at Rum Jungle South was completed in 1963. The milling of the uranium continued until 1971 and, following the completion of the CDA contract, went straight into the AAEC's stockpiles. In November 1960 an officer of the Northern Territory Administration reported that 'trees along the banks of one stream are dying and water holes [are] devoid of fish'. In March 1962 a senior engineer reported to the Administration that severe pollution existed for eight to sixteen kilometres down the East Finniss River from the uranium mill. In January 1963 it was reported that 'there have been heavy concentrations of pollution . . . as large numbers of fresh water shrimp . . . and small fish resembling herring have been floating or lying on the banks.' In June the Report of the Senate Select Committee on Water Pollution contained the following statement:

One of the major pollution problems in the Northern Territory is that caused by copper and uranium mining at Rum Jungle. The strongly acidic effluent from the treatment plant flows via the East Finniss River into the Finniss River, making the water unsuitable for either stock or human consumption for a distance of twenty river miles. Vegetation on the river banks has been destroyed and it will be many years before this area can sustain growth.

On 13 May 1971 a Northern Territory Administration team visited the area and reported that 'no significant rehabilitation has been carried out'.

The main cause of the pollution was badly designed tailings dams. The dams, which were meant to prevent acidic materials and heavy metals used in the milling process from reaching rivers and streams, frequently overflowed during the wet season. As a water resources technical officer reported to his superiors in April 1965, 'the worst period for pollution of the river has usually been immediately after the breaching of the wall holding back the effluent at the treatment plant'. In fact, in the early period of operation, there was not even a dam wall to contain the tailings, which were simply discharged onto a flat plain and allowed to drain into the river. Successive walls were then built and washed away by floods until, in 1961, the tailings were discharged into disused open-cuts rather than onto the flat plain.

In 1970 and 1971 the degree of pollution at Rum Jungle was causing great concern among officers of the Northern Territory Administration and Opposition members of Parliament. The Senate Committee Report referred to above presumably caused considerable embarrassment to the AAEC and generally helped to highlight the problem. In March and September 1971 two interdepartmental meetings took place which revealed the Commission's obstructionist tactics. The first of these was held at Electricity House in Sydney on 24 March 1971 and was attended by representatives of the AAEC, the Departments of Supply, Treasury, National Development, the Interior and the Northern Territory Administration. The minutes of the meeting show clearly a marked difference of opinion between the AAEC and the Department of the Interior on the extent and seriousness of pollution.

The AAEC's Dr R. Warner referred to it as a 'minor local pollution problem'. 'The East Finniss River is not a dead, highly polluted river stream...' he said emphatically. Warner further illustrated the AAEC's lack of genuine concern by stating that, 'the AAEC is still looking at ways to further improve ... [this] minor ... problem and a few more measures may be taken in the short time remaining' (before the Commission left the area). A senior member of the Department of the Interior, Mr Petit, took Dr Warner to task. He said that in general he saw the pollution 'as a significant, substantial problem'. Moreover, he argued for the public release of a study carried out by the AAEC on water pollution in the area.

At this time a major point of contention was the conditions under which the AAEC should be allowed to divest itself of responsibility for Rum Jungle and to hand over the role of caretaker to the Northern Territory Administration. Various government departments were clearly concerned that the AAEC intended to leave the area with a major pollution problem which, by its own carelessness, it had created. The AAEC representatives at the meeting advised Mr Petit that the AAEC would take all necessary measures for pollution control before it left the area, only if this did not require AAEC officers to be stationed at Rum Jungle. This seemed to clarify the AAEC's priorities.

The minutes of a top-level meeting held on 24 September 1971 in Darwin are even more revealing. The meeting was chaired by Mr I. S. Watson, the Acting Director of the Water Resources Branch of the Northern Territory Administration, and was attended by Dr Warner of the AAEC, the Deputy Administrator of the Northern Territory, and eight other scientific specialists (in various fields) employed by the Administration. The meeting was notable for the obvious frustration and anger felt by the Northern Territory representatives at the failure of the AAEC to cooperate in discovering the causes and remedies (if any) for the pollution. Throughout the meeting Dr Warner evaded questions and made assertions that were immediately contradicted by one or more of the others present. He also made it quite clear that the AAEC was not seriously interested in further attempts to revegetate the denuded landscape. Perhaps most revealing was his refusal to make available to these public service colleagues a report on water pollution at Rum Jungle.

During the 1969/70 wet season, the NTA supplied detailed data, in the form of water samples, to the AAEC. The AAEC, which had requested the data, appointed an industrial chemist (Dr Lowson) to analyse the data and write a report. The departmental representatives wanted to see the results of their samples for two important reasons. One was that it would assist them in formulating standards for future mining in the region. The other was that they were responsible for monitoring contaminants in the food chain. Because heavy metals and radioactive materials can concentrate up to one million times as they move up the chain, such pollution poses a grave danger to health. The report by Dr Lowson would have contained information on this subject. The report, though completed, has never been made public. The AAEC has even refused to show it to other government departments.

The whole unfortunate episode, which has environmental effects that will continue for a long time also reveals a certain degree of complicity between government and companies. In June 1971, Mr R. F. Felgenner, First Assistant Secretary, Northern Territory Economic Affairs, presented to his Minister a submission in which he sought approval to investigate the situation at Rum Jungle. In the submission he said:

Early in 1962 the Minister for Territories informed the Minister for National Development that, while the source of pollution has been established beyond doubt and constituted an offence against the provisions of the Control of Waters Ordinance, he was reluctant to proceed against the companies for reasons of their association with the Commonwealth in the venture.

The Minister for National Development replied that the AAEC would minimize the possibility of pollution, but unfortunately any attempt to overcome the pollution hazard would involve quite unreasonable operating costs.

An assessment of the current situation was given in a paper written and published by the AAEC in September 1975. Although the source is questionable, it is the best we have. The report noted that both the East Finniss and the main Finniss rivers were affected. The East Finniss is devoid of fish and plant life; only sparse vegetation lines its banks. At the end of the dry season, concentrations of heavy metals are 'very high' as is the acidity of the water. Great doubt exists about both the ultimate fate and the impact of the large quantities of heavy metals (including radium) that have flowed down the two rivers. It is known that about 2300 tonnes of manganese, 1300 tonnes of copper, 200 tonnes of zinc, and 450 curies of radium have been released from the mine and mill. Radium is a highly toxic substance. When studies were carried out on the radium watch-dial painters in the 1920s, it was found that 0.4 microcuries (or four millionths of a curie) in the human body was sufficient to cause cancer. The radium produced by the Rum Jungle mine and mill therefore contained around 1000 million human cancer doses. About one quarter of this radium found its way into the Finniss River and probably to the sea.

According to the AAEC the variable course of the Finniss in its lower reaches makes it 'difficult to predict just where the released metals may have gone': about a hundred square kilometres of floodplain were affected by the discharges, and average concentrations of copper, manganese and zinc are 1500 per cent, 1400 per cent and 33 per cent above natural levels. The situation with regard to radium is very uncertain. Whilst only a few per cent of the amount released is contained in the surface soil of the area, the whereabouts of the remainder is not known. As the AAEC said rather vaguely, it 'has been removed elsewhere, has migrated through the soil profile or, less probably, has yet to reach the plain'. Though the location of the radium is uncertain, the concentrations in the sampled area are known. At the time of sampling, the radium levels in food, soil and water in the sampled area were slightly above the level recommended as a maximum by the International Commission on Radiological Protection. As part of this was comprised of natural levels, it must be compared with the level in nearby areas (given that natural levels in the sampled area are not known). As the natural level in the area of the proposed Ranger mine is only slightly lower, there is probably little cause for alarm at present. But, because of doubts about the location of the radium released, the future could hold unfortunate surprises.

At present, therefore, if the AAEC is to be believed, the major problem is severe heavy-metal pollution over a wide area. The fact that these metals can be concentrated one million times as they move up the food chain gives added significance to the rather horrifying figures for concentrations noted above. Moreover, the AAEC's reputation for excessive secrecy and highly biased presentation of information on nuclear matters, must cause scepticism about their latest findings: the situation might very well be worse than the AAEC will reveal. But even the AAEC admits that heavy metals will continue to be released from the tailings for up to a hundred years. They will then remain in the environment for a much longer period.

Late in 1975 the Ranger Inquiry visited Rum Jungle and took note of the serious pollution of the East Finniss River and the surrounding area. The Ranger Company and the AAEC have since provided masses of data to 'prove' that the Rum Jungle experience will not be repeated. This data was examined in detail by the Commission, which recommended, should mining go ahead, measures far more stringent than company proposals.

Certainly environmental awareness has increased since the tailings dams at Rum Jungle were designed and built in the early fifties. Nevertheless, the same interests are involved – companies concerned primarily with profit, and the AAEC, which vigorously promotes nuclear power. Both are desperately attempting to assure the public that uranium mining will not have significantly adverse effects on either the local or the global environment. It would be naive to accept their assurances without question.

In order to protect the Alligator Rivers area from the kind of desecration practised at Rum Jungle, the creation of a national park has been proposed. In its publication 'Notes on a Proposal for a National Park in the Alligator Rivers Area' (1972), the Northern Territory Reserves Board introduces itself as the body administering national parks in the Territory – and follows with the cryptic remark, 'There are not many'. In 1965 the Board wrote to the Administrator of the Northern Territory, seeking his approval, in principle, for the establishment of a national park to consist of most of the land between the East and South Alligator Rivers and bounded on the east by the Arnhem Land Reserve. This area was well over 6000 square kilometres. The Board stated that it was anxious to acquire, while it was still available, a large tract of country in the Northern Territory.

The subsequent history of the park reflects that of most other potential wilderness reservations throughout Australia: the exclusion from the reservation of all areas of proven or possible commercial value. In April 1966, Senator Gorton, representing the Minister for Territories, said 'The government is sympathetic to the creation of more national parks but in this case the reservation is complicated by an Aboriginal Reserve, a Wildlife Sanctuary, special purpose leases and pastoral and mining activities in the area. No decision will be possible until these have been investigated thoroughly.' By July 1966 the Reserves Board feared that all suitable land would be alienated before the establishment of a national park could be effected. The board submitted an alternative proposal for an area of about 2000 square kilometres, one third of their original proposal. In April 1967 the Reserves Board again submitted the original proposal supported by a lengthy submission, but 'in desperation' submitted alternatives for a park greatly reduced in area. In January 1968 the Acting Administrator recommended a reservation of about 2500 square kilometres, stating that there were 'no barriers to this reservation.' Mr Sam Weems, Parks Adviser to the US Department of the Interior, touring the area in November 1968, stated that 'The Northern Territory could be the site for one of the best national wildlife parks in the world ... There is no time to waste in getting this area set aside.'

However, concessions continued to be made to commercial interests. In January 1969 a pastoral lease was granted over the Mudginberri area of about 1100 square kilometres, and at Munmarlary of about 1000 square kilometres. In March 1969 the mining firm of Noranda Australia was given permission to prospect. Similar concessions were given to Air Navigators Pty Ltd in August 1969 and to Geopeko and the Project Development Corporation in November 1969. By August 1970 the Reserves Board was extremely worried that the area had still not been reserved and that no authority administered it. The Chairman wrote to the Administrator, enclosing a written report giving details of mining operations. He stated:

My Board has been pressing for the reservation and dedication of this area since 1964 and, for this reason, board members felt an obligation to draw this recent development to your attention, particularly as no other authority seems concerned in this matter. The Board's fear is that once Geopeko are established many others will flock to this area and this will sound the death knell of any plan to preserve this magnificent area for posterity.

This plea fell upon deaf ears: nothing was done.

In July 1971 Noranda Australia applied for five mineral leases close to Nourlangie Rock. The applications were withdrawn after the Reserves Board, Northern Territory Museums and Art Galleries Board and the Australian Conservation Foundation lodged objections. At the same time as Noranda applied for its mineral leases, the Minister for the Interior said that the establishment of a national park was at least two years off because of 'some interesting discoveries of uranium and so on'. In October that year the government decided

Figure 2.2 The diminished Alligator Rivers (Kakadu) National Park, Northern Territory

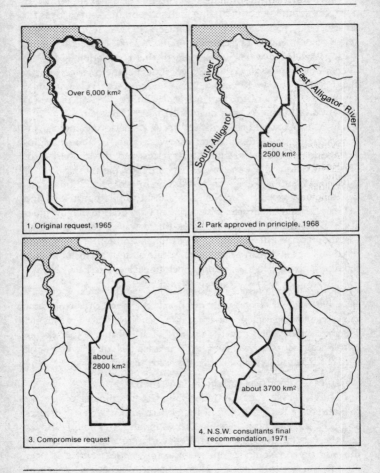

Source: 'Proposal for a National Park in the Alligator Rivers Region', Australian Conservation Foundation, 1975

Figure 2.3 Alligator Rivers Park proposed by environmentalists

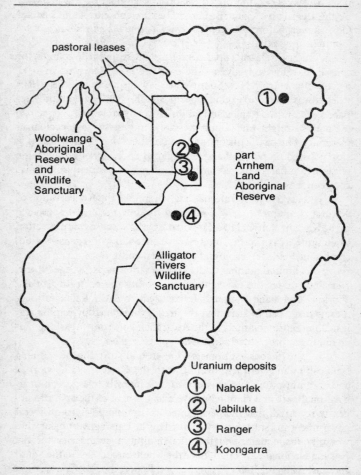

Source: From a proposal made at a national symposium of environmental groups in Darwin, July 1975

to renew all prospecting authorizations. The Northern Territory Reserves Board ends its history of the Kakadu National Park by stating that 'In May 1972, the bill to establish a national park was passed by the Legislative Council. It requires the consent of the Federal Government before it becomes law. *The consent will not be given.*' [our italics]

The Reserves Board's prediction has, to date, proved accurate, for consent has not been given. The park proposal drawn up in 1971 by consultants from New South Wales lay dormant until early 1974 when it was altered to exclude specifically the Jabiluka uranium site. Barely evading the Ranger and Jabiluka sites by a few hundred metres, the park's straight-line boundaries are a travesty of sane ecological planning. The park's history and present boundaries show that it is clearly based on political and commercial criteria, not on the value of the area as part of the national heritage, nor on basic environmental common sense.

In January 1971 the International Union for the Conservation of Natural Resources (IUCN), confirmed the international importance of the Kakadu National Park. It said the park, 'will provide protection to a unique area of special interest to science and world conservation, combining wildlife, scenic and Aboriginal cultural values'. The area is one of a limited number under consideration as part of the 'World Heritage', under the Convention concerning the protection of the World Cultural and National Heritage, which Australia has ratified. IUCN is disturbed to learn that the area is threatened by mining and is making representations to the Australian government, asking that no mining or other development take place in the area.

In July 1975 conservationists from around Australia, attending a symposium and field trip dealing with the Kakadu National Park and uranium-mining issues, decided unanimously that dedication as a national park and comprehensive management of the area was urgently required, that the park boundaries proposed by the Liberal government and substantially adopted by the Labor government were woefully inadequate, and that the minimum boundaries for the Kakadu National Park ought to be the whole catchments of the South and East Alligator Rivers. In its Second Report the Fox Commission recommended that a major national park should be established in the region, including at least one large total river catchment.

Uranium Mining in the Alligator Rivers Region

Two hundred and twenty kilometres east of Darwin and rising nearly 300 metres above the Magela plains stands, at the heart of the Alligator Rivers region, symbolic Mount Brockman. The traditional Aboriginal owners of the land believe that if that land is disturbed, the striated face of Mount Brockman will weep tears of grief, warning of universal catastrophe. At its feet lies the Ranger uranium site.

The region comprises the catchment of the South and East Alligator Rivers. To the south and the east lies a rugged, largely untrafficable plateau. Its steep edges form the spectacular Arnhem Land escarpment that rises up to 250 metres above the undulating plains of the lowlands. Deep dissections and horizontally bedded rocks form a rugged terrain of bare rocks and sparse vegetation. Coloured cliff faces, massive boulders, crevices, gorges, waterfalls, springs, streams, rockpools, caves, sandy fans and distinctive vegetation types give the region a unique scientific and aesthetic value. To the west lie the lowlands, rolling plains with some rocky hills and ridges. Drainage from the lowlands goes either directly into the South or East Alligator Rivers or to extensive floodplain areas lying between them. The lowlands provide diverse habitats for a number of large and small mammals among extensive eucalypt woodlands, tall open forest, savannah, grasslands, scrub, and fringing communities along watercourses. It is in these lowlands that the major deposits of uranium are found.

The floodplains of the region are sections of coastal plains inundated annually by fresh water. They begin where the Nourlangie and Magela Creek systems leave the lowland country and extend east and west to the Alligator Rivers, along the rivers in wide bands and north to the tidal flats of the river estuaries. These plains, when flooded, form the majestic Woolwonga and Magela wetlands. The vegetation of the floodplains is composed of sedges, herbaceous swamps, grasslands and almost pure stands of paperbarks. In addition to these, a variety of fringing communities with pandanus and freshwater mangrove occurs around lagoons or in patches of thin lines in various parts of the plains. In the dry season the floodplains are the refuge of many species of aquatic birds which inhabit swamp areas in very large numbers. To experience the first rays of sunlight rising over Mount Brockman or the dance of a thousand native birds over a flooded billabong is to experience but a small part of the rare beauty of the Alligator Rivers region. The rich diversity of habitat, ranging from rocky escarpment country to pockets of lush monsoon rainforest, makes the Alligator Rivers region a delicate and irreplaceable heritage. This diversity gives rise to a rich fauna and flora. About fifty-one species of native mammals, seventy-five of reptiles and at least 250 of birds, are known. Among the plant groups are scarce relic communities of dense evergreen, non-eucalypt rainforest and of semi-deciduous forest, survivors from another age and climate.

The area is also one of prime archaeological importance. The report of the Archaeological Survey of the Alligator Rivers Environmental Fact-Finding Study reported in their general conclusions that:

The Alligator Rivers region is one of the most archaeologically significant areas in Australia. Research carried out in the vicinity of the East Alligator River in the mid-1960s substantially contributed to the development of a comprehensive Australian prehistory.

This then is the ground under which lie the richest deposits of Australian uranium. The proposed mining operations would be among the largest in the world; the Ranger mine could produce 3000 tonnes per year. Already the area bears the scars of the exploration process, which, if mining goes ahead, would spread over and essentially destroy the ecology of this uniquely beautiful and important region, as the following description of current and projected mining activity shows beyond any doubt.

After the anomalies have been mapped out by radiometric surveys from the air, ground crews carry out more detailed surveys, trekking all over the countryside. Sometimes helicopters are used to reach more inaccessible areas. As they do so, Aboriginal elders wait in the bush to see whether the geological crews will land on their sacred sites. Normally the orebody is mapped out in the Mines Department office in Darwin. Then comes the test drilling, using diamond-tipped drills on the back of trucks or percussion-type drills, to determine the extent and grade of ore beneath the ground. In the Northern Territory uranium fields the water table (the upper surface of underground water usually supplied by rainfall, seepage from the surrounding hills and from nearby rivers) is very close to the surface. Swamps occur where the water table reaches the surface. As exploration causes physical changes to the surface topography it will undoubtedly alter the normal flows of groundwater around the area. The land is usually riddled with drill holes before an accurate estimate can be made as to whether or not the orebody is economic. The drill holes are plugged to prevent entry of foreign materials in case the hole needs to be deepened at a later date. Long trenches about half a metre wide carve up the land as the companies try to gain more accurate measurements for reports to shareholders' meetings. Drill rigs are now a common sight in the Territory where once no white people ventured. Exploration camps and airstrips to enable the supply of materials during the wet season have been established. Roadways dissect the bushland, crossing many of the 'dreaming paths' that have such deep significance to the Aborigines. There has been no attempt to restrict the building of roads to areas agreed to by the Aboriginal people. As much as 5.75 million dollars has been spent on the development of the Ranger uranium deposit since its discovery, a fact that the Company now uses as justification for opening its mine. The problems of erosion of the land and the erosion of Aboriginal culture by the transient drillers and surveyors have already had an effect. (Geopeko was forced to dismiss a number of employees when they caused trouble on Aboriginal land.)

This is the extent of damage so far to an area that many believe should be protected completely from the intrusion of European mining companies. But it is a mere scratch on the surface compared with the results of full-scale mining. If mining goes ahead this is what would happen to the Alligator Rivers region.

During the construction phase it has been estimated that a workforce of 600 would be brought in, the major part being contract labour from the southern states. As the Ranger First Report says, this number of men would not be significant in solving unemployment problems in the south, but such an invasion of the wilderness area in the Territory would make a significant environmental impact. The workers would be housed in a temporary township of parallel roads and small boxes of houses (air conditioned for the executives) a little suburbia standing where once a rich wildlife roamed freely in an open terrain.

Most of the operations planned for the Northern Territory would be large open-cut mining, with the exception of some underground shaft mining, if it is allowed. Depending on the size of the orebody, a huge array of heavy earth-moving equipment would launch a full-scale mechanical attack on the land covering the orebody. First the vegetation would no doubt be burned and then bulldozed. This permanent scar would be scoured deeper and deeper as the waste rock overburden was cleared to expose the ore. Much of this soil and rock would be used for construction purposes such as earth embankments, dam walls and so on. As the removal of any material is so costly, the engineers who plan the construction phase attempt to maximize the use of any earth that is moved. In the case of the Ranger project it has already been suggested, in evidence to the Inquiry, that 'borrow pits' may be necessary. These 'borrow pits' – yet more holes – provide additional material for use in construction.

An explosives store would be built, as blasting would be used in the massive excavation of the open-cut pit. Then the earth-moving equipment would move in to dig out the uranium. Each day, 4100 tonnes of ore and 13 500 tonnes of waste material would be mined. Since the water table is so near to the surface in this area, problems with water inflow into the excavations would be encountered almost immediately, especially when the wet season occurs during the construction phase. Pumps would be needed, working incessantly to keep the workings clear of water. These operations would affect the equilibrium of water movements underground in a way that is difficult to predict.

The hole would grow deeper, until it became a cone-shaped pit 680 metres in diameter at the surface and 175 metres deep, terraced with seven-metre-high steps wide enough to accommodate the large earth-moving equipment spiralling down to its bottom. During excavation by blasting away the sides of the pit, dust and radon gas released from the uranium must be allowed to settle and disperse before the rock debris is trucked away.

The next stage is the processing of the ore. Since the uranium ores being mined in Australia are generally of a low grade and each deposit has its own special requirements for extraction, the individual companies need to build their own processing plants. They have looked at the possibility of a common processing and treatment plant but, because of competition between them, have been unable to reach agreement. Uranium ore needs to be crushed and purified. The uranium is normally processed into uranium oxide ('yellowcake') before it is exported. The ore grade to be processed at Ranger is 0.25 per cent uranium oxide, with the possibility of recovering uranium from ore grade 0.02 per cent uranium oxide by 'heap leaching' using barren acid solutions from the treatment plant. Anything below these concentrations would be discarded. The rock would be put through three crushers, each of 25 000 tonnes capacity. A water-spraying system would 'minimize the ore dust and radon' hazard to workers. The crushed ore would be deposited in towering piles and then be put through the tertiary crusher. Here, if the dust collectors were to fail, workers might be affected by radon gas. Collectors and scrubbers, no matter how regular the servicing, are prone to prolonged failure because repairs take time. It would be tempting for an unregulated company to continue operations, nevertheless, to avoid the expense a shutdown would entail. After passing through the tertiary crusher, conveyor belts would feed the ore into the processing plant. Large bins would maintain the flow to the rod and ball mill, as fluctuations would make processing difficult.

In processing, sulphuric acid is used to extract uranium from the crushed ore. Raw materials would be shipped in to manufacture sulphuric acid at a plant on the site. This acid plant would require smoke and fume stacks designed to disperse sulphuric oxide contaminants as widely as possible. This pollutant has the effect of scorching leaves, altering alkalinity and the chemical balance of soils. Even minor changes of this nature to the sensitive environment of the region would have a radical effect. Moreover, other acidic wastes from this whole operation need to be neutralized. This in turn would require a limestone supply, probably to be quarried somewhere in the Territory. Limestone mining in itself has long been a hard-fought conservation issue in Australia and some opposition could be expected once the companies announced which area was to be quarried. Another material required for the processing would be pyrolusite iron, which would have to be shipped in. The result overall would not be just a hole in the ground, but a major industrial complex, polluting the area with a number of chemicals and shattering the peace of the surrounding country with the incessant noise of the crushers. This could be multiplied a number of times over should all the companies involved commence mining and set up their own processing plants.

The dust and rock residue of the ore form the tailings, a grey sandy material when dry. The tailings, along with the mill water, would be stored in a tailings dam, contained by the earth embankment progressively built up as operations continued. The dam would be subject to the usual engineering problems, such as piping and cracking, which could result in overtopping and uncontrolled release of contaminants onto the flood plain. The tailings dam would also be subject to earth tremors and the vibrations produced from blasting in the pit. These all add to the possibility of failure of the dam and catastrophic release of contaminants into the environment. (For more detail on the content and effect of tailings, see Chapters 1 and 4.)

The dam has been designed to release over 900 000 litres of contaminants by seepage through its bottom. Among the contaminants would be acids and heavy metals like uranium, copper, lead, salts of manganese and the radioactive daughter products of uranium. These contaminants would filter through the groundwater, some being attached to clay particles on the way. The acidity of the water increases the solubility of these metals, making it easier for them to enter food chains. Radionuclides and heavy metals have been found to concentrate as they pass up the food chains; in other words, as these substances pass into plants and then animals and then humans, they are 'collected' so that a regular diet of even small quantities of a contaminated food can reach the level of a dangerous dose. Aboriginal inhabitants still consume bush food and would be extremely susceptible to residual poisoning. During the wet season the excess mill water and tailings would be routinely released with no regard for the effects of concentration when the waterways diminish into billabongs during the dry season. The flushing action of the Magela Creek is prevented by the numerous embayments in its catchment. The effect of incoming tides apparently contributes to the flooding on the plains. Therefore any contaminants released from this or any other mining operation on the Magela Creek would not be rendered harmless by dilution, as the Ranger Company asserts in its Environmental Impact Statement. The effects of flooding around the proposed site have been considered and some modifications of design made to the plant, but there is always a chance that extremes of climate can defeat the purpose of engineering design. Some of the contaminants in the tailings will remain toxic for many thousands of years. In their final submission to the Ranger Inquiry, the company has admitted that,

'The tailings dam area should remain a restricted area to the public due to radiation levels.' Yet they have not demonstrated how this will be possible over a long period on land that the Aboriginal people regard as their own.

In order to prevent the escape of radioactive contaminants, the Company proposes eventually to revegetate the tailings dam. But experimental work done so far is insufficient to guarantee success. The contaminants within the dam could affect the growth of any plants to be used in revegetation. The Aborigines have called on the Company to restore the area to its natural state. But for the revegetation of the tailings dam with species of plants native to the area, it would be necessary to cover the dam with one to two metres of non-toxic fill and a substantial surface of soil on top of that. The Company has not included this proposal in their submissions to the Inquiry. In any case, the companies in general have not considered specific methods of revegetation – the word in just a vague term used by them to quieten opposition. Without guarantees, or an independent body to monitor any discharges, it is impossible for anyone to believe honestly that the effects of the operation will be minimal. All the studies undertaken so far appear to address themselves to engineering feasibility rather than biological reality. To be able to design a system that minimizes effects on the biological systems, it is necessary to have sufficient knowledge of these systems. Zoologists from the AAEC admitted to the Ranger Inquiry that only after a great deal of work has been done can it be known for certain that widespread despoliation of the environment, due to accidental or planned release of waste water, will not take place.

Not all the waste from uranium mining would go into the tailings dam. On the plain, a hundred-metre hill of waste rock would stand, rivalling Mount Brockman, containing concentrations of sulphides that stimulate bacterial action to produce acid wastes. Amid the sad remnants of a landscape that took millenia to create, it would stand, a monument to a brief, twenty-year venture. The whole operation would take place in an area subject to extremes of climate. These extremes could render worthless the most careful of engineers' designs. Sudden inversions of temperature could result in conditions where, far from being dispersed, sulphuric and radioactive gases would accumulate round the mine site and proposed regional centre.

The Second Fox Report concluded, 'That the Ranger project as proposed, and in the land use setting which was assumed [should] not be allowed to proceed' (p. 335). It recommended that an impervious blanket be used to prevent seepage from the tailings dam and that eventually all tailings should be returned to the mine pits. 'This recommendation is based largely on the evidence pointing to possible long-term adverse ecological effects due to continuing seepage losses from the dam, to doubts about the integrity of the dam over centuries and to the problem of radon emission from the tailings if they were not submerged' (pp. 159-60). The Report also proposed a Standards and Monitoring Control Committee, independent of the companies and the AAEC. The latter is described as, 'an active commercial and political force in the promotion of nuclear development and the mining of uranium' (p. 350) and as such, unfit to supervise the mining operation. The Commissioners had visited Rum Jungle and seen for themselves the results of AAEC mismanagement. Most important of all, the Second Report recognized that sufficient knowledge about the region was not available. A research institute should be set up to co-ordinate research and monitoring programmes and that mining should not start until *'it is demonstrated that all components of the monitoring system operate satisfactorily'* [our italics]. (p. 332)

The Ranger proposal includes the development of a regional centre large enough to service 20 000 people. This influx of Europeans and European lifestyle into the area would obviously make a major impact on the local environment and especially on the Aboriginal people. The town would suffer from all the usual problems of remote mining settlements: an imbalance of the sexes, heavy consumption of alcohol, etc. - problems that would, no doubt, especially affect the Aboriginal people, as they themselves fully realize. The Second Report stated that,

The arrival of large numbers of white people in the region will potentially be very damaging to the welfare and interests of the Aboriginal people there. All the expert evidence on this matter was to the effect that, despite sometimes sincere and dedicated effort on the part of all concerned to avoid such results, the rapid development of a European community within, or adjacent to, an Aboriginal traditional society has in the past always caused the breakdown of the traditional culture and the generation of intense social and psychological stresses within the Aborigines. (pp. 232-3)

Figure 2.4 Proposed Ranger mine and treatment plant

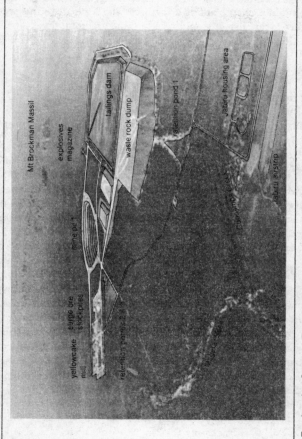

Source: Fox Inquiry, Second Report, p. 13

It is not possible, therefore, to overestimate the impact of uranium mining on the Northern Territory. Into a priceless area of natural and archaeological value would be imported the toxic influence of industrialization. It is not just a matter of digging a hole, getting the uranium out, then planting a few trees and leaving. In the twenty to thirty years that it would take to mine uranium, the ecology of the Alligator Rivers region would suffer a major and irreversible blow.

The Aboriginal View

The Northern Territory is not, as is often thought, an 'empty' wilderness: people have lived there for a very long time. The local Aboriginal population has lived in harmony with the Alligator Rivers region for 25 000 years. These people have a relationship with nature that few Europeans will ever attain. The Berndts, in their book *Man, Land and Myth - in North Australia: The Gunwinggu People*, write:

> The Gunwinggu and their neighbours shared a closeness with nature and a direct dependence on it - a preoccupation that is reflected not only in their mythology, but also in ritual activity. They saw themselves as an integral part of their physical environment, not set apart from all other living things within it but having an intimate relationship with them.

This relationship is the essence of Aboriginal myth, an expression not of a dead past, but of a living tradition. The Berndts say,

> The mythical era in Western Arnhem Land was one of discovery and consolidation. All the main characters [of the myths] helped in one way or another to prepare the land for human habitation. They supplied natural resources, including food, and through their actions in life or memorials left after their death they introduced changes in the landscape - in its rocks, watercourses and vegetation and so on. Above all, they left the spiritual essence which makes the land alive. In this sense they are referred to as 'djang'.
>
> To Western Arnhem Landers, the land is not inanimate and unresponsive like a thing or an object. Its topographical features are a record of 'who was here and did what', for anyone who has the verbal keys to it. Especially, the record tells 'who is here now'. A djang is, by definition, immanent and relevant to the contemporary scene.

This passage expresses the full significance of the Djang - sacred

Figure 2.5 Aboriginal interests and land claims

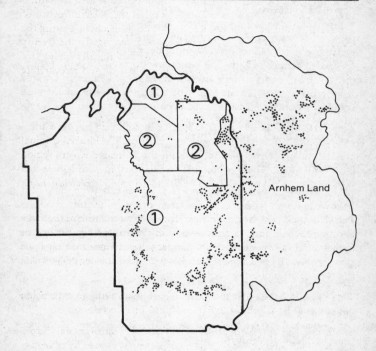

- Aboriginal sacred and archaeological sites
- ① Aboriginal land claim submitted to Ranger Uranium Environmental Inquiry
- ② Additional claim made in February 1977

Source: Based on a submission to the Fox Inquiry by G. Chaloupka, 'The succession of land owning clans to the territories in western Arnhem Land', December 1976

sites – to the Aboriginal people. All the known uranium deposits lie close to such sites. What guarantee can the mining companies give that these places will not be desecrated? And what is the guarantee worth if the spirit of it is ignored? A photograph of the Nabalco bauxite operation at Gove during construction shows a banyan tree, sacred to the Yirrkalla people, standing alone on a desolate flat amidst tin sheds, mounds of earth and heavy machinery. This type of treatment seems to epitomize the attitude of the mining industry towards Aboriginal culture, an attitude which led the Australian Mining Industry Council to place advertisements in major Australian newspapers openly seeking public support for a campaign attacking the Aboriginal Land Rights (Northern Territory) Bill, 1976.

At Gabo Djang, the Aboriginals believe that the one-centimetre-long insects that give the site its name are descendants of the Great Green Ant. They revere the great ant as one of the beings that established all the patterns of human life. They also believe that if the hallowed ground is desecrated the green ants will turn into monsters that will ravage the world. If the proposal to mine the uranium goes ahead the Aboriginals' fear of an ensuing catastrophe could well be justified. In the words of Vai Stanton, whose tribe lived near the devastation at Rum Jungle:

The Oenpelli people warned against disturbing the sacred areas, one of which is the Green Ant Dreaming Place and the eggs which are represented by the large stones in the area.

People were told that if these eggs were disturbed, everyone would be punished and the people destroyed.

They say that the egg place was disturbed and the devastating Cyclone Tracy was only one sign of it.

The Ranger site, about 320 kilometres to the east of Rum Jungle, is the traditional tribal land of Aborigines who are dispersed over the Northern Territory. As part of the Ranger Inquiry, these traditional owners were taken to inspect the site. They were struck by the proximity of the proposed mining operations to their sacred dreaming place, the Djidbi-Djidbi, the Rainbow Serpent dreaming and Dabji. These dreaming places are set in the face of Mount Brockman. The legends surrounding these places tell of catastrophes that will follow if any of the rocks are disturbed. The Ranger Company (who incidentally claim to be the first arrivals in the area, ignoring 25 000 years of Aboriginal occupation) had mapped out an anomaly right at the foot of Mount Brockman, the anomaly apparently disappearing under the foot of the rock debris that had fallen from the top of the massif. Some discussion of the boundaries proposed by the Company had taken place before inspection by the Aborigines. During these discussions, the Aboriginal custodian of the sacred site was eventually coerced into deciding their fate. With every meeting between the mining company and the custodian, the negotiated lease boundary crept closer to the sacred sites.

The way in which these discussions took place was a matter of much concern to the Inquiry. It became clear the the mining company wanted to include the orebody within their lease boundary. The company representative, Mr McIntosh, told the Inquiry, 'There is the question of this No. 2 anomaly, right down near the fence, and it is quite definite that there is uranium in that [area of the anomaly] which is north of the line, and I think the company would be reluctant to give up a claim to that particular thing at any time from now on' (transcript, p. 326).

During the Ranger Inquiry site inspection, much tension was evident in the Aborigines as they unloaded from their vehicles onto the land that the mining company had decided to tear up in search for uranium. They had only a few days earlier observed, at Rum Jungle, the devastation of which mining companies are capable. Concerned feelings emerged about how close and how susceptible to desecration the sacred sites really were, for the mining operations would be only a matter of a few hundred metres away. There was a very heated discussion among the elders and the community, to the point where one of their leaders was heard to say to one of the Commissioners that 'the boundary was too close'. Asked how far away it should be, the response was that it should be about five kilometres north of its present position. If this were done, the boundary fence would cut through the major part of the Company's main No. 1 orebody and part of their site for a processing plant and tailings dam. This was one of the first public expressions of concern for traditional title to the land by Aborigines, and the company men began to look worried.

The following day hearings were held under some trees at the Mudginberri cattle station which was situated downstream of the proposed

Ranger site and in close proximity to the Pancontinental exploration camp. Many Aborigines had gathered for this important meeting, having been told of a white judge who was to look into the uranium-mining question. Judge Fox (as he was known to them) had already flown over the area involved and had a meeting with some of the Aboriginal people before the hearings commenced. Serious attempts were made initially without success to design a proper procedure for taking evidence from a completely different culture. Common law courtroom procedure and many of the legal etiquettes had to be abandoned for they would have provided an inadequate means for expression of the Aboriginal point of view. It was necessary, therefore, to hold another set of hearings in Darwin later on.

The hearings at Mudginberri lasted two days and revealed that there was much confusion among the Aborigines as to what the uranium mining would in fact lead. The Aboriginal experience with mining operations was minimal. There was still much to be learned before any decisions could be expected. The sections of the Arnhem Land Aboriginal communities represented at those hearings expressed concern for their 'bush tucker' (food derived from the land). Aboriginal witnesses attempted to refute the suggestion that they were interested in the dubious benefits of employment opportunities at the mine site. On the second day of hearings, a statement was read on behalf of a meeting of Aborigines held that same morning. Apparently they were most disturbed at the previous day's hearings, which were carried out with scant regard for Aboriginal meeting procedures. Instead a courtroom situation (primitive in surroundings only) had developed, presided over by three white Commissioners and a panel of advisers, all white, including the adviser on Aboriginal affairs. The absence of a black Australian on a Commission to inquire into mining on Aboriginal land began to appear a glaring omission. Feeling was so intense that the chairman of the Oenpelli Council, after the morning meeting, refused to attend the hearings in the afternoon.

The next day the hearings moved to Gove. Planes were hurriedly chartered for those who could afford them. This excluded the environmentalists who were thereby denied representation at an important session of the Inquiry where a head-on clash between mining and Aboriginal interests took place. Nor were the Yirrkalla tribe,

Uranium Mining in Australia 55

whose survival depends on the land and seas affected by the Nabalco bauxite mining operations at Gove, adequately represented.

Realizing that the hearings with the Aborigines were by no means fair, the Commission made desperate attempts to inform them what the mining proposals would lead to. The Commissioners prepared a statement containing an expression of their view of the evidence presented so far. This statement was translated into the Gunwinggu language and circulated as a cassette tape-recording among as many Arnhem Land people as it was possible to reach. Those who claimed some tie or link with the land in question were asked numerous questions and the answers were recorded. This was transcribed and the notes were accepted as evidence at the second Ranger Inquiry hearings held in Darwin. The mining companies objected strongly to the tape, claiming bias and over-emphasis on the hazards and dangers of uranium mining. The Commission was forced to reiterate that the statement was merely an expression of views on evidence tendered so far.

Both the attempt to set up a dialogue by this means with the Aboriginal people, and the resulting transcript of the tape, are remarkable and unprecedented. The tape gives a graphic account of what uranium mining would mean to the people and environment of the Northern Territory. This is the transcript[1]:

This is what I'll be talking about: what they will be doing at Jabiru Mining Company, called Ranger MC, and government. (Not ranger who looks after animals but Ranger MC.) I will be talking about what Ranger will be doing at Jabiru.

What will they do with that place?

They want to dig to get the uranium they have found, then they will send it to different countries.

How many European people will be working there?

Six hundred will come to get the place ready, clear the place, build houses, etc. Six hundred – mostly single men. They will take three years to get the place ready.

That number 600 – how many is that, a lot?

Yes. The same as all the people at Oenpelli and outstations.

A lot.

That 600 who get the place ready, put up factories, etc., when they've finished in three years, who will work then to get the uranium (to sell)?

Two hundred and fifty Europeans will come to dig for uranium. That's as many people as the men and women at Oenpelli.

When those men dig, what will the hole be like?

They will dig two very big holes. First of all they dig out all the rock on top of the uranium. They dig deep; it will be very big. The rock they dig out will be put on a big heap, rock and earth making a hill. Then the stuff underneath with uranium in it they will blast with dynamite and gelignite. A big truck, when it goes down to the bottom of the hole, will look, when we stand on the top and look down, like a toy one. The hole is very deep and wide like a big billabong.

That big hole you talked about – what will happen about the *djang* [sacred sites] nearby?

They've put a fence and they say no one will go past the fence. The holes will be between Jabiru [where the buildings are now] and the sacred sites.

You've said there's uranium. I've seen that place. It looks like rock. The uranium is in it. How do they get the uranium? [What do they do to get it out?]

They dig. The rock without uranium they put in piles and make hills [of rock and soil]. They go down deep and they blast the stuff that they know has uranium in it. Big trucks work all the week and go down and take out the rock with uranium in it and take it to the factory and crush it. They use acid to get the uranium out. The uranium they put in big drums and close them and take it by truck to Darwin. The stuff left over they put in a big dam that they will build. That's what they'll do.

That hole, you said, they put the left-over stuff in – what will the hole be like?

It is a big wide hole. They will make concrete walls as high as the T & G building in Darwin.

The acid they make to get the uranium out – what do they make it from?

They will bring the stuff they need from Darwin in big trucks. They will come every day, many trucks. The road will get busy. They will make the acid and it will get the uranium from the rock. They say – the government and the mining company – they say they will look after the place so it will not be like Rum Jungle.

When I went to Rum Jungle I saw that the trees were burnt, what are they going to do at Jabiru?

They say they will look after it. The stuff left over after getting the uranium, they say, they will put in the dam and it will not spill out, they say. They won't let the stuff spill over and spoil the area. They say. And they say Jabiru country is different to Rum Jungle. You know tin we mine, that we wash and get the tin out, it doesn't hurt our bodies. But uranium is different: it has radon. The people who work there will have special clothes and masks. They'll be looked after. They'll look after the creeks as well and animals so it won't kill them, and they say they will be careful so it won't hurt us. We can't see what uranium has [gas] but if it comes into our bodies it will make us sick or dead and if it's in the water we drink it's dangerous. They say they will look after this so it will be all right.

How many years will they be mining for? A few years?

They will take twenty years to get all the uranium, or thirty years if it's a bit more.

How long is thirty years and twenty years?

A long time. We'll be old. The children who are babies will be grown up; some will be married.

The three holes you talked about. When they've finished getting uranium, will they put the rock and dirt back (the hills)?

No. The holes they will get the uranium from (two), they will leave and they say they will become like lakes used for boats or swimming [i.e., very deep]. The other hole I talked about, they say, will dry out and trees and grass should grow. They will look after it. That's what they say. The hills made of rock, etc., they will leave as hills.

In that twenty or thirty years will the place become a big town?

Yes. They'll build houses for married people and families and single men, it will become big. If the other mines start at same time it will become very big indeed. They'll be building hotels, picture theatre, school, hospital, many shops, sports field, airstrip will be bigger, tourists will come, it will be a big place. It will become like Alice Springs. Alice Springs will be smaller. Jabiru will be bigger if other mining companies are working too.

Are they already mining there?

Not yet, they are still arguing about it. The government wants to mine to send it to other countries to get big money to help run the country. And the mining company who'll be doing the work will get big money too. Some people, both Aboriginal and European, are arguing about it. Some say that uranium is no good. It is dangerous (can kill) us people and animals, and they say that the countries we sell it to may make bombs to kill us. That's why some people (Aboriginal and European) don't like mining for uranium. No one will be able to see what the other countries use the uranium for. Some people say there will be a lot of dust and smoke from digging, trucks, factories, etc. Animals and fish might die (frightened by noise) and water spoilt and uranium will kill animals, trees, fish. It will spoil it all. That's what some people are frightened of.

Is it dangerous? Is that why they are arguing?

Yes. It's dangerous, it can kill us, animals and country (spoil them) and that's what some people say. We don't know, and they are frightened; they [other countries] may make very dangerous bombs that will kill us. Some people who have heard about this mining are upset; they don't want mining near the reserves and parks. They don't want the mining to spoil the country and they don't want lots of tourists coming to spoil it. Some say that it's near where Aboriginal people live. Our rock paintings may be destroyed. A lot of tourists may come, our sacred sites may be spoilt. That's why some people don't want mining.

Jabiru. What are we going to do? Are they going to take it from us? Is it for Europeans?

We don't know. They are still arguing, some don't want the mining and some people want it. Three men in charge [Fox Commission] are waiting for word. They'll listen to the people who want mining, and say they'll be careful, and the people who don't want it. These three men will listen to what's said and if they feel it's all right they'll say to go ahead with the mining, and if they hear it's not all right they'll say not to go ahead. We too – they are waiting for word from us. That's why they sent the paper to us to make this tape. They want word from us. What do we want? Do we want them to mine or not?

It may be dangerous and kill us.

Yes, we don't know. It may be dangerous. Some have already said so, that's why they are arguing. Let us wait for the decision of the three men. These three men will decide. If it's all right to mine they'll say mine, if they feel its no good then they'll say not to mine. If they say to mine then we Aboriginal people can say what we don't like. We can if we don't want the stuff from the tailings dam to go and spoil the water where we hunt, or poison the water we might drink. We can say if we want them to fill in the holes when they finish. We can ask them to prevent too many Europeans coming in and making trouble; we can ask them to look after rock paintings and sacred sites, and not to kill [hunt] the animals and leave them. That's what we can tell these three men, and they take our word and make laws telling Europeans not to go and make trouble, not to come into Aboriginal communities (country). No Aborigines will look after our country [no authority].

As I said, if they mine we can ask for laws to be made on what we want, like Europeans not coming in, not to spoil our paintings, when they finish we want the holes filled in. That's what we can tell them. The other things [we can tell them], if they don't mine, the law is made strong. Well that's all right if we don't want it. But if the others beat us and they say, yes you can mine, well they'll mine and we can tell them what we want them to do to look after it.

This that we've talked about is for us Gunwinggu people, Jim Jim etc., north, south and west. If you want us to help each other, well let's stand up when they make the meeting in two months. These three men will make meeting and we can go and talk about this country. They sent us the paper and we've made a tape for us Gunwinggu people. This is not just fun, this is very important. So listen, and let's work out what we want to say and some can go and say if we want mining or not.

In another attempt to hold proper and fair hearings of the Aboriginal viewpoint, the Commission held private meetings with the representatives of interested parties. One major difficulty was to persuade Aborigines to face up to the uranium question. It was clear that some of them wanted nothing to do with the subject. But willy nilly, the need for making a decision had been thrust on them by the intrusion of the mining companies and the federal government. An immense responsibility had been placed on the shoulders of Australia's original inhabitants who, few in number and subject to the blandishments of mining companies and a multiplicity of government officials, suddenly found themselves the focus of an issue of ultimate international implications.

However, at the Darwin hearing of May 1976 which followed, the Aboriginal case was strongly put. Statements were delivered in a packed courtroom where nervous mining company representatives were outnumbered by the black Australians, who had so much to lose if mining went ahead, supported by a few environmentalists sitting on the floor. The statements, the result of many meetings between Aboriginal legal counsel and the traditional owners of the land in question, were slowly read out by Silas Roberts, Chairman of the Northern Lands Council. The broken air conditioning, along with the content of the statements, drew beads of sweat on the brows of the inappropriately dressed company men. The statements summarized the position thus: 'The act of mining cuts across Aboriginal traditions . . . How can we who follow Aboriginal traditions say we will allow anything that hurts our land?'

At the time the whole question of Aboriginal land rights was before Parliament. The Commission announced at a three-day session of hearings in Sydney in December 1976 that it would return once again to Darwin in February 1977 to examine the Aboriginal position under the new land rights legislation, so new that no one at those hearings was quite sure whether the Act had received the Royal Assent or not. The Act gives traditional owners of the land extensive rights, but prevents an ultimate veto on mining by an 'overriding national interest' clause.

What will happen in the end to the Aboriginal people of the Northern Territory and their uranium-bearing land is still an open question.

The Oenpelli fight on to preserve their traditional values and culture, though with little hope of the support of white Australians. The final submission to the Ranger Inquiry of the Oenpelli Council, prepared after many meetings with the assistance of a draft submission by their counsel, Mr E. C. E. Pratt, but written independently, accepted the inevitability of mining, but reluctantly bowed to it as an intolerable oppression of the Aboriginal people. The submission said,

Aboriginal people at Oenpelli object to and are resentful about the continually increasing number of Europeans that are entering the Arnhem Land Reserve. Some have permits, some do not. Those with permits are often involved in a European-oriented project that may not have been opposed by Aborigines, but at the same time, does not have their support.

In relation to mineral exploration, most of the companies working in the area first applied to the appropriate department, which in some cases consulted Aboriginal groups, who in any case were aware that they had no power to oppose exploration. All companies presently operating in the area were already operating at the time of the announcement of the Whitlam government's policy of land rights and have continued to do so.

This attitude of objection and resentment has been presented to the Commission by Aboriginal witnesses when they have referred to Europeans as being 'pushy'. For example, Mr Buranali: 'they get upset because too much pressure' and 'too much pushing people – asking for mining – asking all the time'. Also Mr Roberts: 'We see white man as always pushing,' and Mrs Maralngurra: 'feels nobody can stop those who have started working.'

The problem faced by Aborigines in relation to the many government departments with an interest in the Reserve has been commented on by the Presiding Commissioner, particularly in relation to employment: 'the great number of government departments that seem to be involved in virtually the same operation'. And it applies generally. 'One of the problems ... the confusion that might arise in the minds of the Aboriginal people, is that there are all sorts of people from different government departments ... going out there and interviewing the people ... one does get the impression rather strongly that they prefer to deal with a limited number of private people rather than a whole variety of them.'

The pressure from mining companies, and having to deal with too many different government departments and agencies are important factors in Aborigines describing Europeans as 'pushy'.

The opposition to mining, and to the increasing number of Europeans in the area has been stated by Mrs Maralngurra. Her statement already quoted indicates that she feels powerless to stop what she does not want. Other Aboriginal witnesses (e.g., Mr Maralngurra) have accepted the inevitability of mining and do not even voice the opposition. Aborigines from Oenpelli, who are the traditional land owners for the Nabarlek area, strongly opposed plans by Queensland Mines to develop the uranium deposit. After some years of opposition they began to realize the ultimate inevitability of mineral development and have negotiated with Queensland Mines Limited, re proposals to develop the Nabarlek orebody, and with other companies for exploration rights over areas of Aboriginal land.

The Oenpelli final submission concludes with a number of proposals to limit the impact of mining on their country and people, but prefaces this compromise with the fundamental statement of the Aboriginal view.

If Oenpelli had the power to make the final decision, it would oppose mining. During Council discussion, the Chairman (Mr S. Maralngurra) stated, 'balanda [white men] push, push, push – soon pubs everywhere and they will kill the race. Look at the Larrykeahs – Darwin is their country and they are living on the tip.'

However, as already mentioned, Aborigines have recognized the inevitability of mining, and also recognize that the Commission may recommend in favour of the Ranger development.

In the event, the Second Fox Report endorsed Aboriginal ownership of the lands that the companies propose to mine. It recommended that Crown lands in the area become Aboriginal lands under the Aboriginal Land Rights Act (1976) and recognized that,

> the traditional owners (the Aborigines) of the Ranger site and the Northern Lands Council (as now constituted) are opposed to the mining of uranium on the site ... some Aborigines had at an earlier stage approved, or at least not disapproved, the proposed development, but it seems likely that they were not fully informed about it as they later became. Traditional consultations had not then taken place and there was a general conviction that opposition was futile. The Aborigines do not have confidence that their own view will prevail; they feel that uranium mining development is almost certain to take place at Jabiru, if not elsewhere in the region as well. They feel that having got so far the white man is not likely to stop ... Having in mind, in particular, the importance to the Aboriginal people of their right of self-determination, it is not in the circumstances possible for us to say that the development would be beneficial to them. (p. 9)

Then follows the paragraph that could be said to summarize the white man's history of domination over indigenous peoples in Australia and elsewhere. It reads,

> There can be no compromise with the Aboriginal position; either it is treated as conclusive or set aside. We are a tribunal of white men and any attempt on our part to state what is a reasonable accommodation of the various claims and interests can be regarded as white men's arrogance, or paternalism. Nevertheless, this is the task we have been set. We hope, and have reason to believe, that the performance of our task will not be seen by Aboriginal people in a racial light at all. That our values are different is not to be denied, but we have nevertheless striven to understand as well as can be done their values and their viewpoint. We have given careful attention to all that has been put before us by them or on their behalf. *In the end, we form the conclusion that their opposition should not be allowed to prevail.* [our italics] (p. 9)

The staggering conclusion of that final sentence (the double meaning of the phrase is intended), written in full knowledge of its implications for the Aborigines, has led to bitter condemnation of the Fox Commission by the Aboriginal people.

Oenpelli alone, outnumbered and even with the new land rights legislation, weak in law, do not have the power of final decision. They expect the pattern of European conquest that has destroyed so many peoples to be repeated, for the profit that uranium would bring to a few. They cannot alone continue to guard their sacred sites that have been preserved century upon century or the natural balance of the land with which they have lived for so long. White Australians can complete, through the mining of uranium, the rape of the Aboriginal people that began with the fatal impact of white settlement, an impact generally deplored in the smart living rooms of Sydney and Melbourne where glossy volumes on Aboriginal culture adorn coffee tables. These books tell of Aboriginal beliefs which say that if their land is disturbed, disaster for the human race will result. The guardians of the 'uranium province' have given fair warning.

3 Reactor Safety
Rob Robotham

In Chapter 1 the nuclear fuel cycle and the hazards associated with it were discussed in some detail, with the exception of reactor safety which was treated somewhat more briefly. It can be argued that in a book about Australia's commitment to the nuclear industry cavalier treatment of this aspect might be expected: after all, there are no immediate plans to build nuclear power plants in Australia. After the collapse of the ill-conceived Jervis Bay project, no State Electricity Authority has done more than keep a watching brief on nuclear power plant development overseas. So why devote a whole chapter to nuclear reactors, reactor safety and associated problems? Principally because one of the main theses of this book is that, although the impact of mining on the Northern Territory is a major concern, Australian uranium cannot – indeed must not – be looked at only in the Australian context. Uranium must be considered in the much broader context of the use to which it is put, and this is supposedly as fuel for nuclear power reactors.

There is considerable controversy over the safety and performance of these technological monsters but before we are in a position to consider these problems we have to see our way through the alphabet soup of reactor types, to try to make some sense of the proliferation of PWR, LWR, CANDU, AGR, LMFBR, etc., and to understand basically how they work. Walter Patterson's book, *Nuclear Power*, gives a comprehensive and detailed account for those who wish to go beyond the outline provided here.

Fission

When a nucleus undergoes fission it splits into two nearly equal parts with the release of several neutrons and a relatively large amount of energy. This energy ultimately appears as heat. This heat can be used to produce steam which then can be used to drive a turbine, thus producing electricity. From the steam production stage there

66 Ground for Concern

Figure 3.6 The nuclear fuel cycle

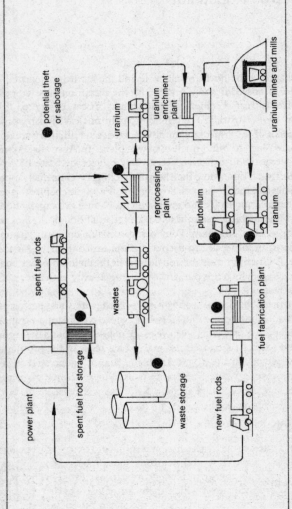

Source: Dale Bridenbaugh

are no differences between nuclear power plants and oil- or coal-fired power stations.

The neutrons ejected following fission are called 'fast' as they are high-energy neutrons. A uranium-235 nucleus struck by a fast neutron may sometimes undergo fission. However, neutrons that have been slowed down or moderated by colliding with other nuclei are much more likely to cause the uranium-235 atom to fission. Such slow neutrons are often called 'thermal' neutrons and reactor types based on fission with slow neutrons are called 'thermal reactors.'

When a thermal neutron interacts with a uranium-238 nucleus, on the other hand, fission is extremely unlikely. More probably the uranium-238 nucleus will capture the neutron and, after two beta decay steps, become plutonium-239. This radionuclide is fissile, that is, it can undergo fission when struck by a neutron. Thus non-fissile uranium-238 can be used to 'breed' fissile plutonium-239.

The fissile nuclei do not always fission into the same products, so when a large number of atoms undergo fission, over a hundred different fission products are formed. Many of these have short half-lives but two of particular long-term importance are strontium-90 (half-life: twenty-eight years) and caesium-137 (half-life: thirty years). A 'half-life' is the period of time within which half the nuclei in a sample of radioactive material undergo decay.

The way the nucleus splits determines how many stray neutrons are given off. For uranium-235 the average is 2.5 per fission. The trick in operating a nuclear reactor is to get just one of these neutrons to produce a further fission, then one neutron from that fission to produce a third and so on. This is called a steady-state chain reaction and is what happens in a smoothly running nuclear reactor. If the 2.5 neutrons produced fissions which then produced about six further fissions which then produced fifteen further fissions and so on, the rate of fissions and the heat output would increase very very rapidly. This is the trick in getting the 'biggest bang for a buck' from a nuclear weapon, getting the fissile material to stay together long enough for the maximum number of fissions to recur. Although working on basically the same principle – nuclear fission – it is happily not possible for a thermal nuclear reactor to become a nuclear bomb. Loss of moderator and neutron reflector prevents such an occurrence. Sadly, the same cannot be said of fast breeder reactors, operating, as might be supposed from the name, on fast neutrons.

Fusion

Before discussing individual reactor types it is perhaps appropriate to point out the differences between fission and fusion. Fusion is effectively the opposite of fission in that two light nuclei come together to form one heavier nucleus.

hydrogen 2 (deuterium) + hydrogen 3 (tritium) ⟶ helium 4 + neutron + energy (lots of it)

It is this fusing of light nuclei with subsequent release of energy that powers the sun and is therefore responsible for life as we know it. However, fusion can take place only under extreme conditions of temperature and pressure, such as are found in the interior of the sun and other stars. These conditions have been reproduced on earth in the hydrogen bomb. Attempts are being made to harness this energy in a controlled way in a fusion reactor but, as the temperatures involved are quite staggering – 100 000 000°C or more – the technical difficulties involved are enormous.

It is anticipated that if ever a fusion reactor works effectively the radiological problems associated with it will be much lower than with fission plants, the main problems being the steady release of tritium and the high levels of induced radioactivity in the fusion reactor structure. However, fusion reactors are several decades away from development so let's get back to the all-too-real world of fission reactors.

Nuclear Reactors

The first nuclear reactor, or self-sustaining chain reaction (known as 'criticality'), was achieved amid tight wartime security in a disused squash court at Stigg Field football stadium at the University of Chicago. Construction started in November 1942 and criticality was achieved on 2 December 1942. Natural uranium metal rods were placed between alternate blocks of graphite, the graphite acting as a neutron moderator. By the time the fifty-seventh layer of graphite bricks had been added it was clear that only the neutron-absorbing cadmium and boron rods were keeping the chain reaction from being self-sustaining. By this time the pile was six metres high and contained 36 tonnes of uranium and over 340 tonnes of graphite. As Enrico

Fermi called out instructions, a young physicist called George Weil slowly removed the final control rod and, at about 2.30 p.m., the world's first man-made, self-sustaining, nuclear chain reaction took place. Reactor safety was not neglected, as a technician stood by with a bucket of neutron-absorbing cadmium salts to dump into the uranium if the neutron count indicated trouble, that was, if the reaction went past the just-self-sustaining stage.

Because of the way the Chicago reactor was built – by piling up graphite blocks – for many years afterwards any nuclear reactor was called a pile. In fact the very first British reactor was called, with atypical British irrationality, BEPO – British Experimental Pile 0 (nought). Since then reactors have been built for experimental purposes, military purposes and latterly for power production. Experimental and research reactors, of which there are two in Australia – HIFAR and MOATA, both at the AAEC's research establishment at Lucas Heights near Sydney – have many uses: to study the effect of neutron irradiation on different materials, to study the reactor physics of different configurations of fuel, and most importantly to produce radio-isotopes for both medical (therapy and diagnosis) and research use. Military reactors have one function – the production of plutonium for the manufacture of nuclear weapons.

All the large early reactors were of this type and, because the heat produced in them went to waste, it was eventually realized that steam and hence electricity might be produced economically from such reactors. Calder Hall in Britain, the first reactor to produce electricity on a commercial scale is principally a plutonium factory. The electricity produced is in effect a by-product. Nowadays, of course, the phrase 'military reactors' could include those used as power plants for both nuclear submarines and nuclear surface ships. These power units were in fact the first power reactors as such, both the USA and USSR starting up such devices in 1954. The US reactor had a power output of 2.4 megawatts whereas the Soviet APS-1 reactor at Obrinsk, now usually accepted as the world's first genuine power reactor, had an output of 5 megawatts.

Commercial nuclear power reactors have one precise and explicit function, to produce useful energy in the form of heat. All currently working commercial power reactors are designed to supply this heat in the form of steam. Thus they have the same function as a gas-

or oil- or coal-fired boiler. Depending on the neutron energy used to produce the chain reaction, reactors are classified as 'thermal' or 'fast'. Thermal reactors usually use fuel consisting of either natural uranium or slightly enriched (up to about two or three per cent) uranium-235. The moderators used to slow the neutrons can be water (ordinary water – H_2O, also known as 'light water'), heavy water (where the heavy isotope of hydrogen, deuterium, is used – D_2O), and graphite. 'Fast' reactors use highly enriched uranium with high concentrations of other fissile material such as plutonium. With a high concentration of fissile nuclei the chain reaction is sustained by fast neutrons. Thus fast reactors do not require moderators.

The core of the reactor is the region where the chain reaction occurs and the heat is generated. In some reactor designs the core is enclosed in a heavy welded steel or pre-stressed concrete pressure vessel. In other designs the core materials are distributed in an array of smaller pressure tubes. The problem is to get the heat out of the reactor core. This is done by a liquid or gaseous coolant circulating throughout the core. Usually the hot coolant passes through a heat exchanger in which water in a separate discrete circuit boils to produce the steam required to drive the electricity turbines. In two reactor types, however, which will be discussed below, the coolant is allowed to boil directly to produce the all-important steam.

The energy output of a reactor can be described in terms of heat as 'megawatts (one million watts) thermal' (MWt) or in terms of electricity produced, 'megawatts electric' (MWe). Since the production of electricity is the main aim of nuclear power, the MWe capacity which is normally only a third or a quarter of the MWt is the crucial factor.

Gas-Cooled Reactors

The British nuclear energy programme has concentrated heavily on gas-cooled reactors derived from the piles originally built at Windscale to produce weapons-grade plutonium. They are known as Magnox reactors because the natural uranium metal fuel is encased in magnesium alloy called magnox. The moderator is graphite and carbon dioxide circulating under pressure is used as a coolant. The British Electricity Generating Boards have built nine Magnox-type reactors and the French, the only other country to opt for gas-cooled reactors, have built seven.

Figure 3.7 Magnox reactor

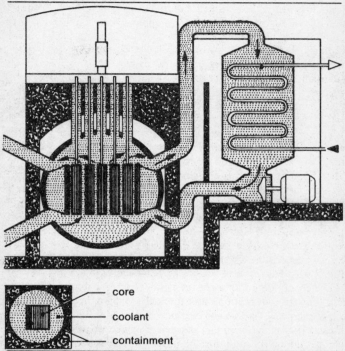

- core
- coolant
- containment

Source: Walter Patterson, *Nuclear Power*, Penguin Books, 1976, p. 51

The basic design of a Magnox-type reactor results in a very bulky construction (Calder Hall, a relatively small reactor weighed 22 000 tonnes) which poses stringent site requirements. Another disadvantage, and a major one, is the low 'burn-up' of magnox fuel. The fuel burn-up is a measure of the amount of energy that can be got out of the fuel and is usually quoted in megawatt-days per tonne of uranium. It is in effect a measure of how many fissions have occurred in a given quantity of uranium. One of the major aims of reactor designs is to achieve as high a burn-up as possible before it becomes too distorted and too full of fission products. With magnox fuel a

Table 3.2 Nuclear power reactor family tree

		Power reactors				
Class	gas cooled		water cooled			fast breeder
Neutrons	slow (thermal)		slow (thermal)			fast
Moderator	graphite		light water	light water	heavy water	none
Coolant	carbon dioxide	helium	light water	light water	heavy water	liquid sodium or sodium potassium mixture
Usual Fuel	natural uranium metal / slightly enriched uranium oxide	*highly enriched uranium oxide	**slightly enriched uranium oxide	**slightly enriched uranium oxide	natural uranium oxide	*uranium plutonium oxide mixture
Reactor Type	magnox / AGR	HTGR	BWR / PWR	SGHWR	PHWR / CANDU	LMFBR

AGR   Advanced gas-cooled reactor
HTGR   High temperature gas-cooled reactor
BWR   Boiling water reactor
PWR   Pressurized water reactor
SGHWR   Steam generating heavy water reactor
PHWR   Pressurized heavy water reactor
CANDU   Canadian deuterium uranium reactor
LMFBR   Liquid metal fast breeder reactor

*Other fuel systems are possible, in particular uranium-233 — thorium-232
**Mixed uranium and plutonium oxides (MOX) could also be used

Source: Ranger Uranium Environmental Inquiry, First Report, 1976

burn-up of between 3000 and 4000 megawatt-days per tonne was the maximum possible. So the hunt was on for an improved system.

Advanced Gas-Cooled Reactors (AGRs)

In effect, AGRs are a second generation, or development of, the Magnox reactor. In the AGR the fuel is enriched to about 2 per cent uranium-235 and is in the form of uranium dioxide pellets, stacked together in their stainless steel 'fuel pins'. The coolant gas emerges from the fuel channels at around 650°C, as opposed to little more than 330°C for Magnox stations. The British AGR programme has run into a whole series of construction difficulties and ten years after the start of Dungeness B, no AGRs are producing electricity.

High-Temperature Gas-Cooled Reactor (HTGR)

Like Magnox reactors, HTGRs are gas-cooled and graphite-moderated, but there the resemblance ends. The fuel is uranium oxide or carbide, enriched up to 93 per cent uranium-235. Thermal efficiency is increased over other gas-cooled reactors as the coolant gas helium is allowed to reach much higher temperatures. HTGR designs being promoted at present allow for a substantial quantity of fertile thorium-232 to be mixed with the uranium fuel, thus producing fissile uranium-233. One advantage of HTGRs is the high burn-up achievable with the fuel: in the international experimental Dragon reactor, up to 100 000 megawatt-days per tonne was obtained. An interesting possibility with this reactor design lies in passing the hot helium directly through a gas turbine, thus eliminating the steam cycle completely.

Water-Cooled Reactors

In these reactors either light water or heavy water is used as a coolant and it usually acts as a moderator as well. Most of the world's present and projected commercial nuclear power plants use light water and are called light water reactors (LWRs). They are of two basic designs, Pressurized Water Reactors (PWRs) and Boiling Water Reactors (BWRs). Like the British gas-cooled reactors developed from plutonium-producing piles, the American PWRs have grown out of the military programme, in this case from the high power density reactors developed for nuclear submarine propulsion. The basic

Figure 3.8 High-temperature gas-cooled reactor (HTGR)

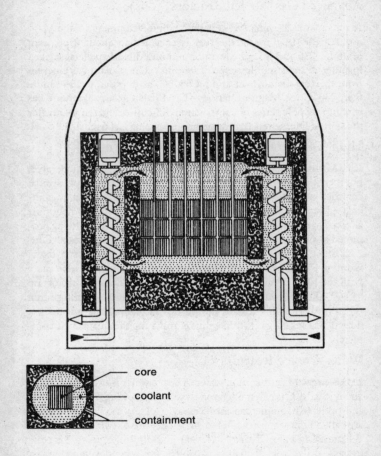

- core
- coolant
- containment

Source: Walter Patterson, *Nuclear Power*, Penguin Books, 1976, p. 60

Figure 3.9 Pressurized water reactor (PWR)

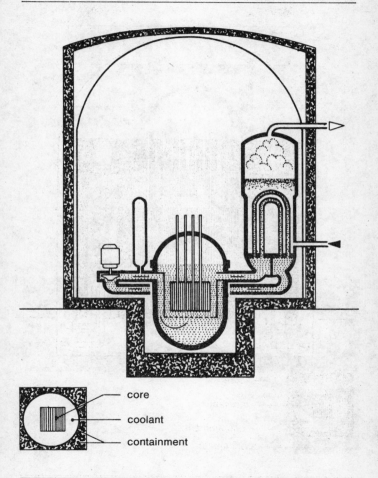

Source: Walter Patterson, *Nuclear Power*, Penguin Books, 1976, p. 64

76 *Ground for Concern*

Figure 3.10 Boiling water reactor (BWR)

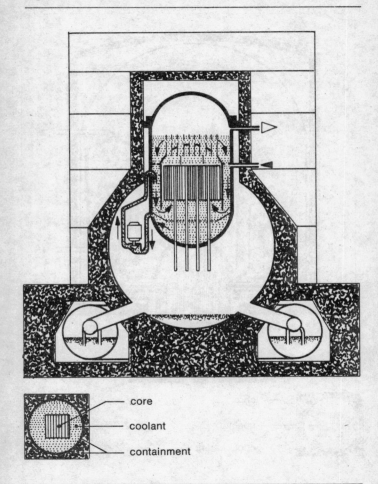

Source: Walter Patterson, *Nuclear Power,* Penguin Books, 1976, p. 68

Figure 3.11 Cross-section of a BWR

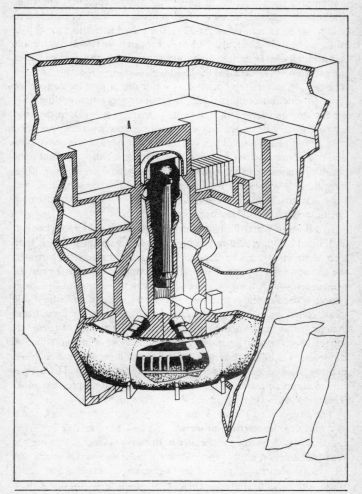

Source: Dale Bridenbaugh

structure of a PWR is a large pressure vessel of welded steel. This contains the reactor core. The remaining volume contains light water under high pressure. The PWR fuel is uranium dioxide, enriched to about 2 to 3 per cent uranium-235, all encased in a thin tube of zirconium alloy called zircaloy. The reactor can have two or more cooling circuit loops. In each loop the pipe through which the water enters the pressure vessel is called the 'cold leg', coming out of the 'hot leg'. The hot leg carries the hot coolant water into a heat exchanger for steam production. Some of the greatest controversy surrounding reactor design has focused on the consequences of a break in the cold leg of a PWR. The emergency core-cooling systems (ECCS) designed to cope with such a break are the most controversial feature of PWR design and are discussed later in this chapter. The low coolant temperature attainable using water under manageable pressure – 150 atmospheres – makes the PWR a relatively inefficient source of heat for electricity production: about two thirds of the core heat has to be dumped in the local environment.

In a Boiling Water Reactor the water serves as moderator, coolant and, in addition, is allowed to boil to produce steam which is then ducted directly to drive a turbogenerator. Once through the turbines, the coolant water is condensed and pumped again through the reactor pressure vessel. A BWR pressure vessel which, because of the lower pressure, need not be as thick as a PWR vessel, must include the whole steam collection and processing array and there are no external steam generators. Since a BWR is coupled directly to the turbine of a generating set, special provision must be made to dispose of unwanted steam if, for any reason, the generator cannot accept it. So a BWR is enclosed in a huge flask-shaped concrete housing called a 'drywell'. Huge pipes lead to a large ring-shaped tunnel called a 'pressure suppression pool'. This steam dump plays a similar role to ECCs in PWRs.

The Pressurized Heavy Water Reactor was developed in Canada as a result of the wartime programme. The Canadian role in fission research was concerned especially with heavy water. Following the post-war decision not to proceed with nuclear weapons, Canada was left with a plentiful supply of its own uranium, expertise in heavy water production, but no enrichment facilities. So the CANDU (Canadian Deuterium Uranium) reactor was devised, using natural uranium as a fuel and pressurized heavy water as moderator and cool-

Figure 3.12 Inside the drywell of a BWR

Source: Dale Bridenbaugh

ant. The heart of the CANDU is a horizontal stainless steel tank called a calendria. The fuel rods, made up of natural uranium oxide in zircaloy tubes, are put into the calendria in an ingenious arrangement designed to achieve optimum moderation and cooling. Four 508-MWe CANDU reactors are operating at Pickering near Toronto, making it about the world's largest nuclear power station. At present, all used fuel from CANDUs is being held in storage ponds but, as the natural uranium fuel design is well suited to plutonium production, no doubt they will be reprocessing eventually.

Figure 3.13 Fast breeder reactor (FBR)

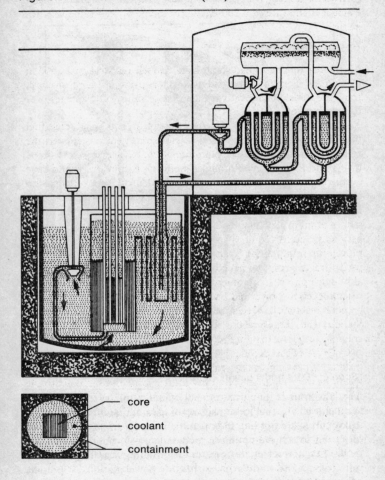

- core
- coolant
- containment

Source: Walter Patterson, *Nuclear Power*, Penguin Books, 1976, p. 78

The British-designed Steam Generating Heavy Water Reactor (SGHWR) is a cross between a BWR and CANDU. It is less thermally efficient than AGR but a 100-MWe prototype proved to be very reliable.

Fast Breeder Reactors

All the reactors described so far have, as their physical basis, fission induced by thermal neutrons. Fast reactors using fast neutrons have some theoretical technical advantages and some disadvantages over thermal reactors.

The major advantage is that they may be able to breed more fuel than they consume. In a thermal reactor some neutrons convert the fertile uranium-238 to fissile plutonium-239. If, for every two fissions that occurred, one new plutonium atom was formed we would have a 'conversion ratio' of 0.5. In a thermal reactor the conversion ratio is always less than one; they are called as a group 'burner reactors'. If the conversion ratio can be made greater than one (and remember that in uranium-235 fission, 2.5 neutrons on average are released per fission, as one is required for further fission, there are 1.5 left for plutonium production). Thus more fuel can be bred than is actually used. Breeder reactors have been developed to the prototype stage although none is in commercial operation. As it happens the first reactor ever to power electricity-generating equipment was a fast breeder reactor. In December 1951 at the Idaho Reactor Testing Station, the Experimental Breeder Reactor-1 (EBR-1) produced enough electricity to light four 25-watt electric light bulbs.

One type of fast breeder is the Liquid Metal Fast Breeder Reactor (LMFBR). To operate successfully using fast neutrons, the reactor core has to be very compact. Not only must it contain no moderator, it must also contain a minimum of other structural material, and only so much coolant as suffices to carry away a fiercely intense heat output, a particularly demanding technical challenge. In the prototypes now operating, liquid sodium is the preferred coolant. This liquid metal is one of the few materials that can cope with the very large quantities of heat produced in the small reactor core. A gas-cooled FBR using helium is under development but, as liquid sodium has a boiling temperature of 990°C, it need not be pressurized – a fact that considerably reduces one major engineering problem. But unfortunately sodium does tend to react somewhat enthusiastically with water, as well as with a wide range of other materials. Thus its open surfaces must be protected from moist air, usually by an inert gas such as argon. This in turn tends to get swept into the flowing sodium, causing bubbles and unwanted perturbations in the cooling circuit. In addition, nowhere must the sodium be allowed to cool below its melting point of 97.5°C or it solidifies, with unhappy results.

Because the sodium can capture neutrons, becoming sodium-24 which emits a particularly penetrating gamma ray, all the cooling circuit has to be shielded. This necessitates a second sodium circuit with a heat exchanger within the biological shield, but shielded from neutrons, before the heat is extracted from the core and taken off to a second heat exchanger for steam production.

Although an exciting concept (breeder reactors could increase the energy obtainable from uranium from about 1 per cent to 60 per cent or more), it does pose horrifying problems since it requires several tonnes of a highly toxic substance – plutonium – at high temperature, cooled by a substance that readily reacts with water. Because of the way the reactor operates, it is possible to hypothesize circumstances under which it could go 'super-critical' – that is, act like a small atomic bomb. One important measure of FBR performance is the 'doubling time', the time taken for the reactor to double the amount of fissile material associated with its operation. Doubling times of the present generation of prototype FBRs is at least twenty years. Designers aim for doubling times of less than ten years, but the engineering required and the implications for safety when working to very close tolerances are quite daunting. One limitation on any expansion of FBR construction is the limited availability of plutonium. To get the plutonium, it has been suggested that the nuclear powers may have to take it out of their nuclear weapons – one of the few arguments in favour of fast breeder reactors.

Reactor Safety

The quantities of radioactive materials present in a nuclear reactor are quite enormous and they must be kept almost entirely out of contact with the environment. The chances of release, in terms of release probabilities, must be kept extremely low. The rest of this chapter examines the dangers and risks from LWRs, as over 90 per cent of the world's proposed reactor programme will be of this type. Furthermore, present and proposed export contracts indicate that Australian uranium will be used in these American-designed reactors.

It is accepted that the maximum credible accident is 'loss of coolant accident' (LOCA). A typical nuclear power plant operates at 600 K. At this temperature the water is at high pressure in water-cooled reactors. A catastrophic accident could take place if a failure occurs in the reactor vessel or steam piping. The water/steam mixture would leave the reactor, an event called blow-down. Even if the break were above the reactor vessel, the superheated water would mostly boil away, leaving the core uncooled. The nuclear reaction would cease as soon as the moderator disappeared but heat from the fission products would be produced at about 6 per cent of the thermal output of the station. If no steps were taken to re-establish core cooling, the core would melt. The subsequent course of a 'melt-down' is unclear. A US Atomic Energy Commission (USAEC) taskforce outlines several possibilities. In a few minutes the molten core would fall to the bottom of the reactor vessel; in half an hour it would melt the fifteen-centimetre-thick steel vessel; in a day or two it would melt through the concrete 'secondary containment vessel'; finally (in about three months) it would solidify about 200 metres or so underground. This is called the 'China Syndrome' in the USA because of the direction in which the molten mass plummets. No one knows what would happen to the radioactive fission products, a quarter of which are gaseous – zenon, krypton and iodine – and would not be contained by the molten uranium core. If cooling is attempted after melt-down has occurred, various mechanisms can be postulated by which the uranium, with the non-gaseous fission products such as strontium-90, could be spattered around the power station and maybe around the countryside as well.

Professor John Holdren of the University of California, Berkeley, has calculated[2] that one quarter of the iodine-131 inventory of a reactor – the amount generally assumed to escape in an uncontained accident – is sufficient to contaminate the atmosphere over the forty-eight co-terminous states of the USA, up to an altitude of 10 000 metres, to twice the Maximum Permissible Concentration (MPC) for that isotope. Half the strontium-90 in a 1000-MWe reactor at the shut-down of the reactor is sufficient to contaminate the annual freshwater run-off of the same area to six times the drinking water MPC. A US study has estimated what could happen under the very worst circumstances – failure of the containment vessel with the release of virtually all the volatile and gaseous fission products into the atmosphere. If this happened in a 1000-MWe reactor situated on the worst possible site in the most adverse conditions of wind and weather, 45 000 to 50 000 people in the nearby population would suffer radiation-induced illness shortly after the accident. Of these, about 3300 would die soon afterwards. About ten years after the accident an increased incidence of cancer would start to occur, with eventually about 45 000 fatalities over the next thirty years or so.

Because of the severity of a LOCA, reactor designers have gone to great lengths to attempt to prevent one happening. As mentioned earlier, the greatest controversy revolves around the effectiveness of what are called emergency core-cooling systems (ECCS). These are the systems designed to operate automatically, and very swiftly, if an LWR primary cooling circuit is depressurized.

PWR Emergency Core-Cooling System

The major function of the emergency core-cooling system of a PWR is to supply sufficient water to cool the core in the event of a break that permits water to leak from the primary system. The break may be very small or a rupture of the largest coolant pipe in the system. PWR emergency core-cooling systems consist of several independent subsystems, each characterized by redundancy of equipment and flow path. This redundancy supposedly assures reliability of operation and continued core cooling in the event of failure of any single component to carry out its design functions.

The first system is an accumulator injection system. It consists of two or more large tanks that are connected through check valves and piping to the main coolant pipes between the primary coolant pumps and the reactor vessel, or to the reactor vessel itself. These tanks contain cool, borated water stored under nitrogen gas at pressures of between about 1300 and 4400 kilopascals (200 to 650 pounds per square inch). If a large break were to occur in the reactor primary system, the hot water should be expelled through the break and the pressure in the primary system should decrease rapidly. When the pressure became less than the gas pressure in the accumulators, the check valves should open to discharge rapidly a large volume of water from the accumulators into the reactor vessel and core. The accumulator injection system is called a passive system because it functions automatically without activation of pumps, motor-driven valves or other equipment.

Two active systems are also incorporated to provide emergency core cooling. One is a low-pressure system for use in the event of large breaks which would result in rapid depressurization and ejection of the primary coolant. This low-pressure system continues to inject emergency core coolant (ECC), from the refuelling water storage tank into the primary system, for a long period after above-mentioned accumulators are empty. The other is a high-pressure system to supply ECC if the break is small and the primary coolant pressure remains high. The active systems have pumps and valves that must operate if the systems are to function. Power to operate the active ECC systems is obtainable from several sources. An active system is energized automatically by coolant pressure and level-sensing and other switches whose operations will connect power (normal or emergency) to the appropriate pumps and valves and thereby initiate emergency core cooling.

BWR Emergency Core-Cooling System

BWRs, like PWRs, have multiple provisions for cooling the core fuel in the event of an unplanned depressurization or loss of coolant from the reactor. The provisions may differ from plant to plant, but all plants have several independent systems to achieve flooding and/or spraying of the reactor core with coolant upon receiving a signal of either high drywell pressure or low reactor-vessel water-level.

Typical emergency core-cooling systems are provided with a high-pressure core spray system (early BWRs) or a high-pressure coolant-injection system (latest BWRs) which are provided to assure adequate cooling of the core in the event of a small leak that results in slow depressurization of the reactor system. If such a leak should occur, the reactor should shut down quickly, on receiving a signal of low water-level in the reactor vessel. Failure of the feedwater pumps (and other minor systems that supply water to the reactor during normal operation) to maintain the water-level above a pre-selected height above the core, or the detection of high-pressure in the drywell containment (outside the reactor vessel), should initiate automatic action to bring the high-pressure injection system into operation. The high-pressure coolant pumps are driven by steam turbines. Steam generated in the reactor (by residual heat in the fuel and in the massive metal of the reactor vessel and internals) drives the turbines. Full pump-speed and flow in twenty-five seconds is claimed for this system.

If for any reason the feedwater pumps and the high-pressure emergency cooling systems should be incapable of maintaining the desired reactor water-level, an automatic depressurization system should operate to discharge steam through pressure-relief valves into the suppression-pool system and thereby lower the pressure in the reactor so that operation of a low-pressure emergency cooling system could be initiated. (The pressure-relief valves should automatically open upon coincident signals of low water-level in the reactor vessel and high pressure in the primary containment.)

A low-pressure core spray system uses two independent loops to provide emergency cooling for use after the reactor has been depressurized. Each loop has two electrically driven (normal and emergency power sources) centrifugal pumps, and each connects through separate piping to a separate spray header above the core. These systems spray water onto the fuel assemblies at flow rates sufficient to cool the core unassisted. Another independent system – the low-pressure coolant injection system – is provided to supplement the low-pressure core spray system and reflood the core. This latter coolant injection system utilizes independent pumps and supposedly has adequate capacity to protect the core following even a large break.

All would be well if these complex emergency core-cooling systems perform as designed. However, by early 1971 a group of engineers and scientists from the Boston area, calling themselves the Union of Concerned Scientists, began to express serious doubts about ECCSs. They drew attention to the fact that there was a total absence of hard experimental data on ECCS performance. The USAEC's own staff, it was subsequently revealed, had been expressing the same concern over this fact.

The ECCS story goes back to mid 1960s and has generated sufficient paper to act as its own biological reactor shield. In 1966 work commenced in the US on a Loss-of-Fluid Test (LOFT) facility. The idea was to design a reactor about one sixtieth the size of a commercial PWR, to initiate a massive leak and see just what happened. This was to take place inside a massive protective containment, since the reactor's demise was expected to be quite spectacular. But the LOFT reactor ran into such a long series of delays (eleven years' gestation so far) and became so expensive that eventually no one could contemplate allowing it to destroy itself.

Thus the designs of emergency core-cooling systems are based almost entirely on computer simulations, the little experimental data available being of questionable quality. The results from the Full-Length Emergency Cooling Heat Transfer experiments, called PWR-FLECHT, and BWR-FLECHT, were hardly encouraging. These tests, carried out on electrically heated, full-size, dummy fuel elements, were dogged by failure – heaters burned out, thermocouples failed to function and even the USAEC's own contractors wrote disgruntled memos pointing out how futile the tests were. In another series of tests in early 1971 on a small model of a PWR, the emergency cooling water failed six times out of six to get into the model core following a simulated pipe break.

The industry's front men, convinced that economic benefits of nuclear power outweigh the risks, set out in late 1972 to show (they hoped) the weakness of their critics' case. The major thrust was the Reactor Safety Study, directed by Professor Norman Rasmussen of the Massachusetts Institute of Technology. A three-million-dollar study, funded by the USAEC, was carried out by AEC staff, consultants and contractors. The report, designated *Wash 1400*, was published in late 1975. In essence its findings indicate that, the more serious a reactor accident, the lower its probability.

The most serious accident, the study envisaged, would be reactor core melt-down (the LOCA) with release of the highly radioactive fission products. The symptoms of the 'China Syndrome' would appear with the core mass acting as a long-term source of radioactive contamination, the effect of which would depend on just where the core settled. The Rasmussen team concluded that there was one chance in 20 000 per reactor per year of this happening.

According to the study, the reactor's containment probably would not be breached and most of the volatile fission products would settle on the cool surfaces inside it. Just a small fraction, with some of the gaseous fission products, would leak out into the atmosphere, mostly within a few hours of the accident. The study concludes that, provided some appropriate emergency measures were instituted, it would be unlikely for anyone to be harmed, but the situation would have to be closely monitored and exposure to the population controlled.

Since its publication the study has been cited time and time again by the proponents of nuclear power because of its optimistic conclusions, but it has also been extensively criticized for the way those conclusions were reached, and not all the criticisms have come from the anti-nuclear power brigade, as we shall see.

The study used the technique of 'fault tree analysis' for determining overall risk. In this form of analysis the effects of a hypothetical accident are traced through the operation of a plant or piece of equipment. The probability of failure and the subsequent effects of the failure are assessed at each stage. To take a very simple example: what is the probability of, say, a bolt holding a pump on a metal frame sheering? If it does, what strains are put on the other bolts? Will they sheer? Under normal operation, or if the pump is overloaded? If they sheer, will the pump break free of its mounting, feed tubes, etc? How vital is the fluid that it is pumping to the process involved? What consequences will the lack of fluid have to the process? And so on and so on.

As Rasmussen explained, 'By use of an appropriate analytical model that includes hardware failures from operating, test and maintenance procedures, it is possible to predict an expected value for the probability of the accident.' This claim is unsupported by the innovators of this technique, the (US) Department of Defence (DOD) and the National Aeronautics and Space Agency (NASA). For instance, in NASA work on the Apollo programme, fault trees, event trees and mathematical models were never put forward as absolute values of system reliability, or as estimates of the time reliability of the system. What fault tree analysis can do is highlight relative probabilities, not absolute probabilities as the Rasmussen team attempted to do.

There is also always the possibility that something no one thought of could go wrong. As the US Environmental Protection Agency, when receiving the study, put it: 'It should be noted that statistical techniques such as this, although appropriate analytical methodology, can never conclusively show that all critical pathways to an accident have been considered.'

The Dresden II incident highlights some of the difficulties of making theoretical predictions about things that do actually happen. Predicting the 'perversity of inanimate objects' is not a rewarding pastime. The Dresden II Boiling Water Reactor went critical for the first time in January 1970. On 5 June 1970 it was undergoing power testing at approximately 75 per cent power. A spurious signal in the reactor pressure-control system altered the steam flow to the turbine and caused a turbine trip (uncontrolled revolution of the turbines) followed by a reactor scram (insertion of the 'scram' control rods to stop fission). Subsequent erratic water-level and pressure control in the reactor vessel, compounded by a stuck indicator pen on a water-level monitor-recorder, and the inability of the isolation condenser to function as needed, led to the discharge of steam and water through safety valves into the reactor drywell. Here the accident sequence itself initiated further malfunctions of safety valves, an example of the 'feedback' that even the most complex models cannot envisage in depth. A reading of the history of the incident, where an operator responds incorrectly to incorrect signals, leading to uncontrollable pressure surges inside the reactor vessel, leads one to agree with W. Castro who, writing in the journal *Nuclear Safety*, stated grimly, 'It is unfortunate that procedural, mechanical and control inadequacies can be recognised only upon the occurrence of some incident that puts them to a real test.'

A possible way of cross-checking the Rasmussen approach is to calculate, using the same sorts of procedures, the probability of those incidents that have actually occurred and which were not foreseen. The Union of Concerned Scientists calculated that the probability of the entire event chain in the Dresden II incident was 2.4×10^{-38} per reactor per year. In effect, this means that either the incident (and fourteen other significant incidents) have never happened, or the fault tree analysis does have certain limitations.

One very important event that Rasmussen deliberately excluded was sabotage and terrorism. He states that 'The study did not explicitly consider the public risk due to potential acts of sabotage.' Elsewhere, Rasmussen has admitted that, 'we do not believe it is poss-

---

*1 chance in 40 000 000 000 000 000 000 000 000 000 000 000 000

ible to meaningfully predict the probability of a sabotage event. This probability varies rapidly with social conditions.'

There are, unfortunately, many examples of terrorist action against reactors. Hijackers of a US Southern Airways flight threatened a US nuclear facility. Scottish nationalists threatened an English reactor. Representatives of the 'People's Pacific Army' threatened French reactors during the 1974 Moruroa atomic tests. Explosives recently damaged a reactor under construction and another operating in France. No doubt terrorists will exhibit a learning curve about the vulnerability of the nuclear fuel cycle in the same way that technologists are claimed to exhibit a learning curve in reactor safety. In wartime, nuclear reactors, reprocessing facilities and waste storage dumps would make attractive targets for an enemy wishing to spread the radiological impact of its weapons.

However, the proponents of nuclear power argue, very persuasively, that there has never been an accident with a commercial nuclear power plant that has resulted in danger to the public. But there have been accidents at military reactors that have posed a threat to the public and there have been some very close calls at commercial power plants. This chapter will look now at only the more interesting ones; it would be impractical, without producing an encyclopaedic tome, to review them all. For instance a USAEC task force reported, for thirty operating LWRs between 1 January 1972 and 31 May 1973, 'approximately 850 abnormal occurrences . . . [which] involved malfunctions or deficiencies associated with safety related equipment . . . Many of the incidents had broad generic applicability and potentially significant consequences.' The task force concluded, 'It is difficult at this time to assign a high degree of confidence to quantification of the level of risk associated with nuclear reactors.'

Canada was the first country to experience a major reactor accident – at the NRX experimental reactor at Chalk River. On 12 December 1952 a technician in the basement of the NRX building mistakenly opened three or four valves (the exact number has never been established), which lifted three or four of the reactor's twelve shut-off rods out of the reactor core. The valves were quickly reset but, due to the wrong information being sent to the control desk, four more shut-off rods were lifted out of the core; the power level of the reactor started to rise so the controller 'scrammed' the reactor. Pressing the scram button should have re-inserted all of the seven or eight rods into the core, but only some went back, and those somewhat slowly. To shut off the fission reaction it became necessary to dump the heavy water, literally a last-ditch action. Unfortunately, the power surge had melted some of the uranium fuel, releasing some fission products. The light water coolant circuit was severely damaged and altogether about 10 000 curies of low-lived fission products, in about 4 million litres of water, leaked into the building basement. The subsequent clean-up was a massive operation. But, considering that there had been an almost complete failure of the scram-rod system, the reactor staff can consider themselves very fortunate.

The NRX was brought back into operation and was able to take over after its successor, the NRU, had to be shut down for decontamination in 1958. An irradiated fuel element broke and caught fire inside the refuelling machine. At one stage a one-metre length of intensely radioactive fuel fell into a maintenance pit, where it burned merrily, contaminating up to 400 000 metres of land around the NRU building.

Britain's turn came in October 1957 when the Windscale plutonium-producing reactor shut down during a routine operation when the physicist in charge allowed the reactor core to overheat. The uranium and graphite caught fire and at the height of the blaze about eleven tonnes of uranium in 150 fuel channels were burning. After all else failed, it was decided to flood the core with water even though there was the possibility that the water and molten metal might react, oxidizing the metal and leaving hydrogen to rise with incoming air and explode. No one could be sure that such an explosion would not rend open the building, releasing the fission products to the local countryside. However, the water worked and the fire slowly died down. At an early stage of the emergency, instruments had been showing radioactivity reaching the filters at the top of the Windscale exhaust stack. These filters were known as 'Cockcroft's Folly'; Sir John Cockcroft, Director of the UK Atomic Energy Research Establishment, had insisted that filters be installed, as a precautionary measure after the stack had been built.

'Cockcroft's Folly' probably kept a major accident from becoming a catastrophe as the filters retained most of the particulate fission products and plutonium. But iodine-131 is gaseous at the temperatures involved and about 20 000 curies passed through the filters.

The critical path of iodine was known: uptake in soil and grass, concentration in milk by grazing cattle, further concentration in human thyroid by people drinking the radioactive milk. Infants and small children are at particular risk after radio-iodine releases because of their large milk intake and small thyroid glands. It was quickly arranged to pour the milk away over an area of 500 square kilometres, giving the local farmers suitable compensation. For a few weeks the milk output of Cumbrian cattle reached amazingly high levels. The half-life of iodine-131 is eight days, so the milk could have been kept for cheese or butter, but presumably the government wanted to avoid any unpleasant public reaction if 'radioactive' milk appeared on the market. Unfortunately there does not appear to have been any systematic follow-up of people who lived in the area at the time of the fire. The locals now talk suspiciously of the number of people who die of diseases like cancer these days, twenty years later. But there is no firm medical evidence either way, because it has not been sought.

After the fire in the number one pile, Windscale II was closed down till after the various inquiries were finished. The full report was never published, but the information that was released made clear that the changes recommended to the second reactor would be prohibitively expensive. Both reactors were more or less concreted up and there they stand on the Cumbrian coast like latter-day pyramids, a monument to man's technological folly.

In the USA there has been a series of accidents. In November 1955, after coolant tests to the EBR-1, the reactor operator mistakenly used slow-acting control instead of scram-rods when shutting the reactor down. The fuel temperature rose to 1100°C and about 50 per cent of the fuel melted. No activity was released to the environment but the core was destroyed. Events like the destruction of the BORAX experimental reactor in 1954, damage to the fuel rods of one of the Hanford plutonium-producing reactors, and fuel-melting in the Heat Transfer Reactor Experiment, the Sodium Reactor Experiment and the Westinghouse Test Reactor, were all relatively minor. But SL-1 changed all that!

A 3-MWe prototype military reactor, the Stationary Low-Power Reactor number 1 (SL-1), was shut down for routine work on instrumentation. For some unknown reason the central control rod, which was disconnected, was pulled out of the core. The result was catastrophic. The core almost instantly went supercritical, the fuel fired itself and the consequent steam explosion blasted what amounted to a solid slug of water to the roof of the reactor. There were three young servicemen working on the reactor. All three were killed, including one who was impaled on the ceiling by an ejected control rod. The three bodies remained so radioactive that they had to be buried in lead-lined coffins in lead-lined vaults.

The Enrico Fermi fast reactor near Detroit was the next to appear in this sorry saga of accidents. In a fast reactor, a core melt-down could result in the fuel being rearranged into what is called a 'fast critical assembly' – a small nuclear bomb, if you like. At a late stage of construction it was decided that this possibility could be avoided by putting a metal pyramid on the floor of the containment so that molten fuel would run off its sides and spread out. The engineers who were building the reactor objected vehemently to this late addition and, as it turned out, they were right. One of the six zirconium triangles forming the pyramid cover broke loose. The liquid-sodium coolant lifted the twenty-centimetre-long triangle up into the core, partially blocking the sodium flow. The temperature of the inadequately cooled fuel pins rose. They melted, distorting other pins, further blocking the sodium flow, in a progressive distortion of the whole assembly. It stopped short of a complete melt-down, but was like the battle of Waterloo – 'a damn close-run thing'.

Mainland Europe has not been without its close calls. The Lucerne gas-cooled reactor in Switzerland had a loss-of-coolant accident after a fuel element burst. Because of the low power density and the fact that it was built in a cavern under a hill, no radioactive materials were released to the outside atmosphere. The Federal Republic of Germany built its first large commercial nuclear power station, the 640-MWe BWR at Wuergassen. It went critical in October 1971 and had its first accident in April 1972. With the reactor operating at 58 per cent of design rating, a pressure relief valve opened and stuck. Over 200 tonnes of steam was condensed in the drywell. The pressure surges tore strengthening structures loose, and leaking water damaged the wiring for the control rod drives. After re-starting in November 1972, it was found that water was again accumulating in the drywell. Investigation showed several short cracks in the primary piping, with one that was twenty-five centimetres long being the actual leak.

But, because of the bizarre way it began, the Browns Ferry fire will take some beating, as it has won nuclear immortality for an old-fashioned candle. On 22 March 1975 two electricians had just finished some cable modifications in the cable spreading room under the reactor control room of the Browns Ferry Boiling Water Reactor, Alabama. They were checking the air flow through the cable penetration with a candle flame when the draught dragged the flame into the foam-plastic packing around the cable trays. The extent of the fire in the cable-spreader room was limited to a metre or so from the penetration, but the presence of the fire on the other side of the wall from the point of ignition was not recognized until significant damage to the cables had occurred. 1225 cables were damaged for Unit One, 64 for Unit Two and 24 for the unfinished Unit Three; 379 were common to all three units and some carried multiple conductors. *Nucleonics Week* reported that, 'the emergency core-cooling system, the reactor core isolation cooling system and remote control for many vital valves of Browns Ferry 1 were knocked out of action by the fire which burned for seven hours before it was brought under control. Both Browns Ferry 1 and 2 were scrammed manually from full power...'

None of the normal or emergency low pressure pumps was working so a makeshift arrangement using a condensate booster was made. The water-level dropped from its normal 508 centimetres above the core down to 122 centimetres, but Unit One did not become uncovered. A similar system worked for Unit Two. There was no meltdown and no release of radioactive material – not this time.

Finally, for those with a taste for black humour, there is the case of the prototype nuclear-powered cargo ship, the *N.S. Matsu*. While on trials, a radiation leak was discovered. Because they were at sea, improvisation was necessary. Firstly, borated boiled rice, then old socks, were used to plug the leak – such is the inexorable march of technological progress. Feelings in its home port of Matsu Bay ran so high that the ship had to drift, like a latter-day Flying Dutchman, for forty-five days before it was allowed to return to Japan.

The industry stresses time and time again that there has never been an accident in a commercial nuclear reactor that has resulted in danger to the public. The SL-1 accident killed three men but they were servicemen, not members of the public. The Enrico Fermi Reactor accident did not release radioactive material off-site, not quite. The Windscale fire did of course release large quantities of radio-iodine but it was not a commercial reactor. Browns Ferry did not melt down, thanks to a jury-rigged pump, and not even the $40 million cost of replacement electricity could be called a danger to the public.

As Walter Patterson puts it, 'The logic of the industry argument is impeccable.' It is to be hoped it stays so. As more and more nuclear electricity gets plugged into the world's power grids, the risk of an accident increases. If there were a major accident (it is tempting to write, 'when') there could well be a massive public revulsion against these technological monsters, with a demand to shut them down. The greater the percentage of energy that is nuclear-generated power, the greater the social disruption that would be caused by taking nuclear power plants out of the grid.

But for all their inbuilt problems and difficulties, the greatest hazard caused by nuclear power plants is the waste they produce, which will be discussed in the next chapter.

4 Nuclear Wastes
Sandy Pulsford

Long after nuclear reactors have ceased to operate, leaving only their empty hulks to litter the countryside, the problem of the radioactive wastes they produced will still be with us. The reverberations of our society's adventure into the atom will continue for a million years and more.

For the first few decades, high-level wastes will be intensely hot from the decay of short-lived fission products. Then for 600-1000 years the longer lived fission products – caesium-137 and strontium-90 – will become the most potent of the isotopes in the wastes. Their place is taken by the transuranium istopes, such as americium-241 and plutonium-239, which will not decay for up to 500 000 years. But longer lived wastes will remain, such as plutonium-242 and neptunium-237, with half-lives of 380 000 years and 2.1 million years respectively, whose lifetime extends beyond 20 million years.

For mankind to create such a legacy for the world, with only a vague notion of how to deal with it, shows, at the very least, a gross arrogance towards all life on this planet. Industrialized societies have exploited, in two centuries, much of the fossil energy accumulated on the earth, over many millions of years. Having used up much of that, we now face the option of substituting, for that energy source, one which imposes, for the next few millions of years, an enormous environmental burden on the earth and countless generations of its inhabitants.

Suggestions abound on what to do with the wastes that pose this threat. Many of the proposals amount to no more than fanciful ideas, such as rocketing them into the sun, or leaving them to melt into the polar icecaps. The continued mention by the nuclear industry of these impracticable remedies serves to point out the paucity of real options open. More seriously considered methods are disposal in the seabed, or in geological formations, and transmutation in reactors. All have their serious drawbacks and will be considered later in detail. None has yet been shown to be acceptable. The commercial development of nuclear power has been built on the industry's expression of faith that a satisfactory means of waste disposal will, one day, be found. This premise is coming under increasing criticism and the search for a solution is starting to take on some sense of urgency.

The debate on nuclear wastes has, for the most part, tended to focus on the high-level wastes produced from the reprocessing of spent reactor fuel. This emphasis is quite justified as these wastes contain something like 99 per cent of the total radioactivity produced in the fuel cycle. Substantial leaks of high-level military wastes from more than twenty of the formerly named US Atomic Energy Commission's (USAEC) 200 mild-steel storage tanks have brought home to people the fragile nature of early storage techniques.

The immediate health risks, however, have tended to come from the other one per cent of radioactivity in lower level wastes arising in other parts of the fuel cycle. The relatively low levels of activity and the diverse range of sources and types have led to a lack of foresight and a lack of care in their handling, creating many serious and potentially serious situations. These range from mining, where hundreds if not thousands of underground miners could lose their lives or die of lung cancer, to the reprocessing of reactor fuel which gives rise to huge quantities of low-level wastes that are discharged to the environment and potentially reconcentrated by it. Many such cases have become issues in their own right and the debate on disposing of all forms of low-level waste has started to increase in intensity over the last few years.

This chapter looks at the range of wastes produced in the fuel cycle and worldwide experience in dealing with them. It then goes on to look at philosophies and issues in the management of long-lived wastes and the options for waste disposal.

Mining

Radioactive wastes are produced at all stages in the nuclear industry. Mining and milling of the ore produces the largest physical quantities of waste in the form of tailings, a fine sand containing the radioactive daughter products from the decay of uranium. Because these daughters have shorter half-lives than uranium, the tailings pile is more radioactive than the original ore and, because of the milling process, the pile is larger than the hole from which it comes. Wind and water erosion can distribute this radioactivity into the environment, increasing the natural background radiation levels to nearby populations. More serious radiation effects occur where radon (the radioactive gas given off from radium) from the tailings is allowed to accumulate, as in thousands of homes and schools in the USA and elsewhere where tailings have been used for building fill and concrete.

The radioactive life of the tailings pile is of the order of 100 000 years. In 1974 the USAEC Directorate of Licensing estimated some 91 000 tonnes of tailings are produced from the annual input to a 1000-MWe Light Water Reactor LWR.[3]

Fuel Fabrication

Enrichment and fabrication of uranium fuel creates only very low-level wastes. However, the fabrication of plutonium fuel elements creates much larger waste problems. At present, no commercial reactor uses plutonium fuel, but plans to introduce the recycling of plutonium in conventional reactors, and to build fast breeder reactors, are expected to increase substantially the problem of plutonium-contaminated solid wastes. These wastes, variously known as 'transuranic wastes', 'alpha-bearing wastes' or more commonly just 'alpha wastes', consist of the vast quantities of equipment and protective clothing that are disposed of daily because of contamination with radioactivity, especially plutonium and other long-lived transuranic elements (the 'actinides').

Such things as disposable gloves, masks, shoes, coats, rags and packaging, as well as tools, gloveboxes, glassware, instruments, chemicals, filter resins, evaporator residues and condensate sludges must be collected for disposal or storage. Much of it is imcompressible and/or incombustible and hence unable to be reduced in volume.

It has been estimated by the Organization for Economic Co-operation and Development Nuclear Energy Agency (OECD/NEA)[4] that some three million cubic metres of transuranic waste will be produced by the US nuclear industry up to the year 2000, arising almost equally from fuel fabrication and reprocessing. This is a much greater volume than that of the high-level wastes anticipated in that time as the transuranic waste could include more than 22 000 kilograms of plutonium, an amount comparable to that lost into high-level wastes. Because of its long lifetime, transuranic waste must remain isolated from the environment with the same safeguards as high-level wastes.

Nuclear Reactors

In reactors themselves, there are many low-level wastes produced which are often not sufficiently considered. Radioactive gases are released from the reactor coolant vents, the liquid waste tank and the equipment vents, and via the ventilator air. Solid wastes, which in normal reactor operations are generally free of actinide contamination, include spent resins, air and liquid filters, evaporator concentrates, laboratory glassware, protective clothing, tools and equipment, and miscellaneous paper and other wastes. Radioactive liquid wastes include reactor coolant coming from expansion during start-up, valve leaks, steam-generator tube leakage and laboratory samples. In addition, floor drains, laundry and decontamination wastes and recycled process water need to be treated for radioactive contamination. These liquid wastes go through various forms of processing to produce relatively contamination-free process water which is released to the environment or recycled in the plant, and a concentrated solid or sludge which is packaged and shipped to land disposal sites or dumped at sea.

Reprocessing

A nuclear reactor's main radioactive effluent is removed from the plant as spent fuel elements. A typical 1000-MWe LWR has a fuel load of some hundred tonnes, of which thirty tonnes are replaced annually. These are stored on-site in pools of water for some time, to allow short-lived radioactivity to decay pending fuel reprocessing.

Reprocessing to recover uranium and plutonium has always been considered an integral part of the fuel cycle. The removal of unfissioned uranium is a basic economy measure, as the fuel is not taken to high burn-up levels, and plutonium is recovered on the assumption that in the near future nuclear power will move onto a plutonium economy footing. This would be based on the use of mixed plutonium oxide and uranium oxide fuel in conventional reactors. Presently, both these options are under serious challenge in many parts of the world. If the plutonium fuel cycle were abandoned, the only justifications for a reprocessing operation would be to produce plutonium for military purposes, or to convert the spent fuel into a more stable form for ultimate disposal.

No commercial oxide-fuel reprocessing plants are in operation anywhere in the world at present, although in a number of countries small pilot plants and facilities from weapons programmes are in operation and many large-scale plants are planned. President Carter's policy to discourage reprocessing is a matter of current international controversy. Much resistance has been shown to it by France, West Germany and Japan in particular.

In 'Purex' reprocessing, the fuel pellets are usually chopped open, releasing tritium and the gaseous fission products krypton and iodine. The remaining fuel is dissolved in nitric acid, from which the uranium and plutonium are then removed by solvent extraction. According to workers at the US Environment Protection Agency (USEPA)[5], the remainder, some 1500 litres per tonne of spent fuel, becomes high-level waste. The residue of plutonium left in these wastes is anything up to 1.5 per cent, but ideally is only 0.5 per cent or one kilogram per reactor-year.

Figures quoted by the Union of Concerned Scientists for the West Valley plant in New York, which closed in 1972, indicate that low-level wastes are also produced in copious quantities – at a rate of between 100 and 10 000 cubic metres of waste – from reprocessing the annual fuel load of one reactor. Equivalent figures from the OECD report[6] in 1971 show the production of up to 15 000 cubic metres. This waste is contaminated predominantly with tritium, which has proved very difficult to remove. The stainless steel or zirconium alloy fuel cladding gives rise to some two cubic metres of highly active solid waste.

Nuclear Facilities

Another major source of waste that is rarely touched upon is the old, disused nuclear facilities themselves. Reactors, reprocessing plants and fabrication plants all have a limited lifetime. This is partly due to the increasing contamination of the plant, rendering it too 'hot' or 'dirty' to allow the necessary maintenance to be done in safety. Dismantling and disposing of many thousands of tonnes of highly radioactive reinforced concrete and steel, to say nothing of trying to bury it a thousand metres underground, is something the nuclear industry will face with some trepidation. Early next century there are expected to be some thousands of defunct facilities needing decommissioning. Western Europe is already worried about finding suitable disposal sites large enough to isolate safely its high-level waste and the even larger volumes of alpha wastes. Effective disposal of decommissioned facilities themselves can only intensify these problems.

Low-level Wastes

Since the days late last century of Madame Curie and the discovery of radioactivity, attitudes to radiation hazards have had to be revised many times. Increasing knowledge and increasing concern for the dangers have caused radiation standards to be tightened by many orders of magnitude. Even now it is admitted that there is not enough known to be sure about present radiation standards and the safety of present practices, and controversy still rages on a number of issues. Every time new standards are laid down or revised, past and present waste disposal techniques are called into question. Here we run into a version of what has been called the 'grandfather syndrome', which poses the question: If a certain method of disposal is found to be unsatisfactory or hazardous, we can improve on our methods of disposal in the future, but what do we do about those wastes already disposed of? For the most part, they are likely to be already dispersed into the environment and effectively beyond recall.

Tailings

The hazards posed by uranium mill tailings are generally ignored by the nuclear industry in any analysis of the routine and accidental impact on the environment from nuclear power. Piles of tailings, in many countries of the world, lie abandoned by the companies that produced them and, until very recently, disregarded (in some cases deliberately) by the respective atomic energy agencies for whom it was mined. These piles of fine sand – some hundred tonnes for each tonne of uranium extracted – are usually uncovered and are often found in or near the mining towns.

Port Hope in Ontario, Canada, is only one example of the costs in financial and health terms of the nuclear industry's negligence in dealing with its wastes. For forty years, mill waste from the production of radium and subsequently uranium was dumped at a number of sites within the town, releasing radium and radon into the soil, water and air around. Not until 1967 was there any thought of the health hazard created, when a nuclear engineering professor measured radium levels at the dump sites and raised the alarm over the high concentrations he found. Some fill was removed and some new dirt spread as a result, but nothing meaningful in terms of a solution eventuated. No surveys of the extent of the contamination were conducted and no monitoring was undertaken until 1975 when radon was found in a local school at concentrations considerably above the recommended maximum level of three picocuries per litre. Eight homes showed radon levels higher than fifty picocuries per litre and were evacuated until remedial work could be done. Of the 433 sites tested in an initial survey, 110 were identified as needing remedial work. Action was only recommended in cases where gamma radiation or radon levels were greater than eleven times the background levels.

Dozens of homes and other buildings had been built on radioactive tailings used as fill. In addition, concrete blocks from a dismantled processing building and laboratories were sold to townspeople. The highest radium concentrations found in the soil were 500 to 1000 times the normal concentrations. The major problem arises when radon produced from decaying radium percolates up through the ground and into buildings above, where it concentrates in the confined space. Radon has been shown capable of moving through concrete floors. Clean-up operations in many houses require replacement of basics such as floor joists and basement flooring. The first stage of the programme involves the evacuation, disposal and replacement of 70 000 cubic metres of material and the temporary removal of twenty-seven families from their houses. However, the clean-up was hampered by residents of nearby towns rejecting proposals to dump the wastes in their areas, which were specifically recommended by a report as the best sites.

This situation parallels, with disturbing closeness, the stories of Salt Lake City, Grand Junction, Colorado and many other mining towns in the western United States. At Grand Junction, radon levels found in one school room were a staggering 300 times the recommended maximum level. It has been estimated that each year residents of contaminated buildings received on average 110 extra millirems of gamma radiation than in background levels and average annual lung doses of five rads from radon daughter products.

Ocean Dumping

The history of dumping radioactive wastes into the sea provides another example of attitudes towards nuclear wastes over the last three decades. The first discharge of wastes was an accidental spill into the Columbia River and thence into the North Pacific as a result of plutonium production operations begun at Hanford, USA, in 1944. The first deliberately introduced material was packaged solid waste deposited into the Pacific by the USA in about 1946, and from 1949 Britain similarly used the Atlantic Ocean. About 1952 the British Windscale works began disposing of waste into coastal waters by pipeline.

It was not until the 1972 London Convention that any effective international agreements were reached on radioactive waste disposal in the sea. The International Atomic Energy Agency (IAEA) is now responsible for permits to discharge wastes.

The Windscale reprocessing plant has the highest discharge limits of any plant in the world which, according to the OECD figures published in 1971[7], allow it to dispose of up to 300 000 curies of beta-emitters each year and 8000 curies of alpha-emitters per year. This implies that, if Windscale discharged its full quota of alpha-emitters as plutonium-239, it would be releasing over 120 kilograms to the environment each year. It is claimed that these limits represent only a fraction of the limiting environmental capacity, as determined by the method of 'critical pathway analysis'. This method identifies the group of people most at risk from any proposed radioactive release and sets waste discharge limits that will keep the dose to this critical group within accepted radiation standards. A number of assumptions underlie this method. The first is that mammals, being the most biologically complex species, are therefore the most susceptible to radiation damage. It would follow that discharge limits set so as to cause no harm to man would not cause damage to marine ecosystems. It does not, however, claim that there would be no damage to individual organisms.

A full proof of this assumption would appear to be a difficult exercise. On the other hand, there is no basis for ruling out a disproof of the assumption, which may show significant ecological damage from waste dumping. We would then be faced with the previously mentioned 'grandfather syndrome' whereby there would be nothing we could do to rectify the situation. The wastes discharged may include isotopes with lifetimes of up to half a million years, during which time food intake habits might be expected to go through substantial changes, leading to different critical groups than those at present, with perhaps much higher intakes of contaminated food. Again, that society, even if it were aware of the situation, could be helpless to rectify it. Critical pathway analysis assumes continued research into identifying critical pathways and critical groups in any given society or cultural situation, for as long as the wastes remain active. This assumption cannot be regarded as realistic.

Two US scientists, Dr Vaughan Bowen and Dr Karl Morgan, have recently challenged ocean disposal of wastes by delineating a new mechanism for the wastes discharged from Windscale to reach man. It was previously assumed that radioactivity was absorbed into sediment on the ocean floor. It now appears from new research that turbulence may cause this contaminated sediment to rise to the surface and be deposited on beaches. From there, it could be picked up by the wind and transported over populated areas. On the ABC's television programme, *Four Corners*, of 12 June 1976, Dr Morgan is reported to have said, 'I'm certain that this would not be acceptable in our country...'

Strong disagreements over standards of dumping still exist between the US Environment Protection Agency (EPA) and the Nuclear Energy Agency of the OECD. The EPA recently investigated ocean dumping sites used for packaged waste in the 1940s and 1950s and found that many drums had begun leaking soon after reaching the sea floor, due to implosion under high water pressure. The USA stopped ocean dumping in 1970 and the EPA now insists that any future wastes dumped should be in containers that will stay intact until the wastes have decayed to innocuous levels – some 300 years. On the other hand, the NEA maintain that solid wastes dumped on the ocean floor would be harmless even if the containment failed completely. The NEA sees the greatest danger as the unintended recovery of the wastes by fishermen or others who may be ignorant of their contents.

A disturbing facet of this study was that, although the locations of the dumping sites and the total quantity of radioactivity dumped was known, many details such as container size, contents, and methods of packaging were incorrect or unknown. It appears that important information had not been passed on over a single generation and within a single stable country. It is hard to imagine the continuity of similar information for centuries, much less millenia. It was on these investigations that the EPA researchers discovered a new genus of giant sponge growing on the broken waste containers, to heights of up to 1.3 metres.

Transuranic Waste

The debate on the handling of low-level and plutonium-contaminated wastes came strongly to the fore in 1976 in the USA and elsewhere, when many government and non-government groups took a fresh look at this area and were not impressed with what they saw. The General Accounting Office of the US Congress found that the half dozen commercial burial sites for the disposal of low-level wastes were chosen for any reason except their suitability for containing the wastes for the necessary period. Reasons were used such as: the land being available on federal and state property; it was not otherwise useful; its placing would attract other nuclear facilities, etc. No consideration was taken of the hydrology or geology of the site. It turned out that the Nuclear Regulatory Commission recently relinquished control of these sites to the state governments who in general have neither the funds nor the expertise to set standards or to enforce them adequately.

The theme of haphazardness and lack of foresight and care recurs frequently in the field of solid waste management and disposal. Much of the solid waste produced in the industry is contaminated with plutonium. Until the last few years, the USA buried most of this in shallow earth trenches up to six or seven metres deep. Radioactive particles were assumed to be absorbed by the soil and fixed there for hundreds and thousands of years until they decayed. Unfortunately, this theory was inadequately tested and large migrations of radioactivity have occurred in the decade or two since their emplacement. Close to a thousand kilograms of plutonium and other long-lived alpha-emitters have been dumped in this way in the USA alone, with no proper safeguards for its isolation. The USAEC in 1974 ordered a stop to ground-burial of commercial solid wastes with more than ten microcuries of transuranic activity per kilogram of waste. The questions remain of how to dispose of future alpha wastes and what to do with the vast quantities of waste already buried. It has already been found necessary to exhume wastes from one trench of low-level military waste at Hanford, Washington, to avoid a possible nuclear chain-reaction from the high concentrations of plutonium.

Another disturbing facet of the debate on low-level wastes is the lack of any agreed international standard for the alpha content of solid waste that would demarcate wastes to be discharged to the environment from alpha wastes requiring storage for millions of years. In contrast to the figure of 10 picocuries per kilogram (pCi/kg) used in the USA, the UK uses a figure of 20 pCi/litre (or 120 pCi/kg) and Japan uses 50 pCi/litre (or some 300 pCi/kg). It seems that the choice of this threshold is rather arbitrary.

The formerly named USAEC operated an eighty-hectare waste burial ground in conjunction with its Savannah River plant. Although this plant has probably been involved more with nuclear weapons than with commercial nuclear power, it must be remembered that it was the nuclear weapons industry that spawned civilian nuclear power and the history of the one is the history of the other. Although the underground water table was only thirteen metres below the surface, solid wastes were dumped here in more than three kilometres of trenches, six metres wide by six deep. Operation of the burial site began in 1953 with transuranic waste ineffectively separated from other waste and relying only on the trenches to contain them. In 1960 plutonium-238 processing began and the activity of the waste increased tenfold. These wastes were placed straight into earth trenches to be covered by 1.2 metres of soil when full. It was not until 1965 that it was decided to allow for the possibility of future retrieval at a later date, by introducing concrete culverts, two metres by two metres, as a primary containment. However, trench burial continued for large pieces of equipment which were encased by pouring concrete over them. This helped to stop radioactivity leaking away, but of course severely hampers future retrievability.

In 1970 storage of wastes in 200-litre drums placed in the concrete culverts was introduced. It was not until late 1974 that the practice of dumping wastes with a lifetime of 500 000 years into the earth was abandoned as just not good enough. Now Savannah River staff are evaluating the possibility of mining, extracting, repacking and disposing of – at a phenomenal cost – some 100 000 m³ of contaminated soil.

The most striking thing about the nuclear industry is the extraordinary ability of people to circumvent its planned mode of operation. Recently, it was discovered that the townspeople of Beatty, Nevada, had been pilfering contaminated materials from waste dumps. Tools, looking brand new after only a few uses, and dumped because of contamination, proved too tempting for the townspeople. Officials also discovered metal waste transport tanks used as septic tanks and water tanks for livestock and humans. After a house-to-house search for contamination, the Nuclear Regulatory Commission took twenty truckloads of stolen materials back to the dump.

As much plutonium is lost into solid wastes as into the high-level wastes. By now, some several tonnes of plutonium have been released to the environment, most of it beyond recall.

Reactor Effluents

Nuclear power plants themselves have consistently run into problems with accidental discharges into the environment. The Vermont Yankee reactor in the USA spilled an estimated 300 000 litres of waste into the Connecticut River on 20 July 1976. The waste contained radioactive tritium and small concentrations of various metals. State emergency services were not notified for a full day after the spill occurred. State health officials said the leak warranted mobilization of emergency procedures.

Sweden's first release above permissible levels came on 18 July 1975 when the Oskarshamn 1 reactor had to shut down as rapidly as possible for an unexplained reason. The back-up cooler had been defective for two weeks and it was known that any emergency shutdown would result in releases of radioactive gas up to fifty times the normal levels. Fourteen days later the utility was still telling visitors that there had never been a release in excess of prescribed limits at a Swedish reactor. The utility had decided the release was not harmful because the wind happened to be blowing out to sea, and it did not inform the State Radiation Institute until the fifteenth day after the incident.

Other releases have occurred from spent fuel storage ponds at a reactor in Florida. The leaks have been continuing since 1974. The Millstone 1 reactor at Waterford, Connecticut, leaked radioactivity from the waste treatment system into the building's heating system, necessitating the evacuation of 150 workers at Millstone 1 and 2000 construction workers who were building Millstone 2. That and a subsequent spill three days later led to dramatic increases in environmental contamination around the plant. Water samples revealed a jump from one pCi/litre to 400-500 pCi/litre in one year. The new World Health Organization standard for limits in water is thirty pCi/litre. Local environment officials said that radioactivity levels had been rising steadily since Millstone 1 had come into service.

Decommissioning of Nuclear Plants

The short experience in decommissioning reactors and reprocessing plants so far has shown that options for dealing with such a mammoth problem are strictly limited. Firstly, the biological shield around the reactor core could be sealed and filled with cement. Secondly, the external structure could be partially collapsed and covered with earth and, thirdly, it could be demolished and the materials removed and disposed of in some way. Estimates of the costs of these operations range from 5 per cent to 100 per cent of the original cost of building the facility.

With the prospect of 2000 reactors and ten large reprocessing plants (with their associated waste storage tanks) by the turn of the century, it is obvious that the policy of abandonment proposed in the first option would quickly become unworkable. It would soon lead to a haunting skyline of dead reactors and the alienation of large tracts of land from all other uses, in many countries. The structure would, in addition, be made virtually impossible to dismantle at some future time. The second option, while perhaps more aesthetically pleasing, could hardly be considered a satisfactory means of disposal. An earth covering could not be regarded as a permanent shield, as it is subject to wind and water erosion. As before, this would mean the alienation of the surrounding land for up to a thousand years. Such a requirement would, of course, be virtually impossible to achieve. The third option would require the demolition of thousands of tonnes of steel and concrete highly contaminated with radioactivity, its reduction to manageable sizes, its transportation and its disposal by some means as yet unknown.

Very little thought has gone into the problems of decommissioning, and numerous defunct facilities, such as the Enrico Fermi fast breeder reactor near Detroit, the two British Windscale reactors, the small reprocessing plant at Mol in Belgium, the larger West Valley reprocessing plant in New York and many other smaller plants, will lie as monuments to the enormous problems still to be solved by the nuclear industry.

Issues of Long-Lived Waste Management

High-level waste from spent-fuel reprocessing represents one of the most potentially hazardous impacts on the environment from the nuclear industry. Many people have found sufficient reason to reject nuclear power solely on the basis that there is no solution to the problems of disposing of radioactive wastes.

From the beginning of the nuclear weapons programme, the industry recognized the need to isolate and control high-level wastes. As an interim measure, military wastes were placed in steel storage tanks surrounded by concrete. This form of storage is still in use today, requiring constant operation of mixers to prevent sedimentation, cooling coils to remove decay heat, and condensing and filtering systems to process the gases and volatile liquids given off. Continual surveillance and monitoring are also required.

Such was the optimism of those developing commercial nuclear power in the 1950s and 1960s that this means of storage was accepted as a satisfactory basis on which to build the industry. The fundamental problems of waste disposal, as opposed to storage, were brushed aside as too hard and as a challenge for future decades to solve. This has remained a dominant philosophy since that time. However, serious mishaps began occurring at major storage areas in the USA from the 1950s onwards, eventually involving more than 10 per cent of the USAEC's 200 tanks. The largest of these was at Hanford, Washington, where, in mid 1973, almost half a million litres of high-level waste leaked into the ground over a seven-week period before the leak was recognized.

It became apparent that liquid wastes stored at or near the surface were very vulnerable to leakage, earthquake and attack, and that costs of maintenance and periodic replacement of the tanks would be considerable in the long term. This led to concepts of extended storage of wastes in a more stable, solid form. It was proposed that this would buy time in the search for a means of waste disposal and allay public concern about the wastes issue. In 1969, regulations were adopted in the USA requiring high-level wastes from reprocessing to be converted to a stable solid within five years and delivered to the USAEC within ten years for extended storage in a proposed Retrievable Surface Storage Facility (RSSF). At that stage the AEC had little idea what sort of facility it was going to build. Work on solidification had, however, been going on for some years.

A number of concepts for RSSF emerged, including a water basis, incorporating forced-circulation water cooling in which 500 steel canisters of waste, three metres by thirty centimetres, would be stored in steel-lined concrete modules. Each module would have its own pump, heat exchangers and cooling tower. Ten canisters, each with a capacity of .177 m^3, would handle the solidified wastes from a one-gigawatt electrical reactor operating for one year (1 GWe = a thousand million watts). This proposal would require an ongoing positive commitment by society, not only to supply energy for the forced cooling system, but also to maintain the equipment, monitor the facilities for leakage and replace canisters corroded by water or otherwise breached.

A second design involved air-cooled casks in which from one to three canisters of the same size as before would be sealed to mild-steel casks with a wall thickness of five to forty centimetres, and surrounded with up to one metre of concrete shielding. These would be placed on individual pads in the ground with convective air circulating between the concrete and the steel casks. This system, while potentially requiring no maintenance – i.e., only a passive commitment by society – has the drawback that, in the event of a breach of the steel containers, there would be no further barriers to the escape of the wastes directly into the atmosphere.

The UK and Canada also opted for building retrievable surface storage facilities for their wastes. The UK, especially, adopted the philosophy that high-level wastes should not be allowed to pass beyond the control of man. This meant having the solidified wastes where they could be kept under constant scrutiny in a man-made storage area. In 1975 Britain officially opted for a water-cooled system similar to that described above.

But the concept of extended surface storage has come under increasing fire from many sources. As far back as 1967 a strong note of warning was sounded on proposals for storage, when the OECD Nuclear Energy Agency, in a paper[8] on ocean dumping of wastes, said:

provision for the ultimate safe disposal of currently produced radioactive waste should be the responsibility of the generation producing it. So-called 'storage' of the waste may, in fact, mean that a final solution to the problem has been postponed . . . storage should not be used merely to pass on unsolved problems to future generations.

More recently, in September 1976, the UK Flowers Report dealing with nuclear power, expressed concern that no real attempt was being made in the UK to look beyond the storage of wastes to some form of ultimate disposal. The report felt that even if an ultimate disposal method were not used immediately, it should still be developed as a back-up, if needed.

In the US proposals for the RSSF were seen as an indication that the industry had given up trying to find a means of permanent disposal. Officials of the US Environment Protection Agency were among the strongest critics. At a conference in 1974 they expressed the fear that, 'the day may dawn, and very quickly, when for economic reasons alone, the "interim" engineered storage facility will become a "permanent" storage facility.'[9] It appears the EPA was afraid that, with development of the RSSF, work towards a means of permanent disposal would be slowed down or stopped and inflation or other factors could then act to make it uneconomic. For example, in a few decades the companies that produced the waste may well be out of existence, leaving an unwilling government to pick up the tab. A similar situation has already occurred in New York, where the operators of the West Valley reprocessing plant decided in 1976 to cut their losses and leave the plant, including tanks of high-level wastes, to the state of New York to clean up.

The USAEC in 1973 did in fact terminate its research in New Mexico where it was investigating sites for permanent disposal of wastes. However, it was forced to resume again two years later under strong public pressure. In the meantime, in September 1974, the AEC released its draft Environmental Impact Statement on radioactive wastes. This was withdrawn for rewriting in early 1975 by the newly formed Energy Research and Development Administration (ERDA), the successor to the AEC. ERDA dropped the idea of extended storage and the RSSF and put its funds back into finding a site for permanent disposal.

Despite these reversals, US experience provides an object lesson in the need for retrievability. Schemes offering retrievability permit undoing a problem that may not have been foreseen, whereas a non-retrievable system gives zero room for error. ERDA's present plans for disposal in bedded salt formations a thousand metres or more underground is clearly a non-retrievable option, as the steel canisters placed in the salt would corrode within a matter of months or, at most, a year or two.

From the mid-1960s, the AEC was investigating a salt mine at Lyons, Kansas, as a possible site for a waste repository. Extensive research and investigations were carried out over a period of years, including the placement of short-lived demonstration wastes. In fact, the AEC had sunk its entire funds for waste disposal into the Lyons site. The pilot project was at an advanced stage in 1972 when two Kansas geologists revealed that a salt mining company that was operating nearby using solution mining techniques had lost some millions of litres of water into the formation. The fate of the water was not known, but the incident cast sufficient doubt on the safety and integrity of the formation that 'Project Salt Vault' was hurriedly scrapped. Successive hopeful disposal sites near Carlsbad, New Mexico, have also had to be rejected by ERDA because closer examination has revealed problems with every one.

The need for safe retrievability must remain one of the absolute criteria for disposal of long-lived radioactive waste, because of the limitations of knowledge about its long-term dangers and the potential for error inherent in any scheme. At the same time, something more advanced and further removed from the biosphere than a surface storage pond must be found to meet the criterion of total isolation of nuclear wastes. The necessity to satisfy both these requirements is something the nuclear industry has not yet faced up to.

The solidification of high-level wastes is periodically offered to the public as a solution to the problems of nuclear waste disposal.

It is claimed that wastes may be 'disposed of' by fusing them into glass. However, it is clearly not a disposal option in its own right, but merely an adjunct to the containment of radioactivity in some storage or disposal operations. Solidification gives a more compact, robust and more easily handled waste product with less chance of releasing radioactivity into the environment in the event of its primary containment being breached. Disadvantages are that most solidification processes are still in the development stage, and present techniques may be superseded in the future. More importantly, until a method of ultimate disposal is chosen and agreed upon, the objectives of solidification cannot be fully defined and the properties required of the solid are not accurately known. If wastes are solidified now, the product may be found unsatisfactory in a few years time. Reprocessing three-tonne blocks of dense, highly radioactive glass would not be easy. Secondly, for solidification to reduce significantly the hazards of storing liquid wastes, it must be carried out immediately after reprocessing, when heat levels are many times higher, giving rise to solid wastes of much greater volume. Otherwise the greater part of cumulative wastes of a nuclear power programme will be in liquid form. It is estimated that, if Britain's reprocessing wastes are left to cool for three years before solidification, then 75 per cent of her wastes will be liquid at any one time. Therefore, any short-term benefit of the solidification programme would be all but negated.

Techniques for solidification are generally in one of two categories. Firstly, some form of calcination – e.g., fluidized bed solidification or pot calcination – in which wastes are heated and gases and liquids are driven off, usually leaving a powder of granules. The calcina is comparatively easy to process further, but has no resistance to leaching by liquids. It is considered unsuitable for anything but very short-term storage, of the order of a decade or two. The second group of techniques involves the fixing of wastes within the structure of a glass or ceramic, up to concentrations of 25 per cent. This is called glassification or vitrification and can reduce the volume of wastes by factors of five to ten.

The stability of vitrified wastes over the required time scale is not yet proven. Accelerator tests, involving the incorporation of high levels of alpha-emitters in the glass, have attempted to simulate the radiation load on the glass over tens of thousands of years. This is still significantly short of the hundreds of thousands of years for which stability is required. West German research on radioactivity in glass revealed problems such as devitrification, local hotspots, increased leachability and surface splintering on contact with water. In addition, if the recycling of plutonium in nuclear reactors goes ahead, the wastes from these reactors will have significantly higher proportions of transuranic elements, increasing substantially the alpha radiation load on the glass. On the other hand, if plutonium is not recycled, then all plutonium produced, and that which is currently stockpiled, will become part of the nuclear wastes, whereas at present only approximately 0.5 per cent to 1.5 per cent ends up in high-level waste. This would increase the alpha load on the glass by one or two orders of magnitude.

Finally, the concept of solidification is generally applied only to high-level wastes. However, plutonium-contaminated solid wastes (alpha wastes) require isolation from the environment for equivalent periods of time and they contain at least as much plutonium as high-level wastes. Solid wastes, contaminated rags, tools, clothing, etc., are typically produced, according to the OECD[10], at a rate of one cubic metre for every kilogram of plutonium processed. Thus the annual output of a 1000-MWe light water reactor would give rise to some 200 cubic metres of plutonium-contaminated solid wastes each year. This is around four times the annual volume of high-level wastes from the reprocessing of spent fuel. Not being homogeneous, these wastes are particularly hard to incorporate into a non-leaching solid, such as glass. Therefore, up to half of the long-lived radioactivity in nuclear wastes is unlikely to be effectively immobilized by vitrification. The short-term benefits of vitrification and of calcination are far from clear cut. In the long term, many questions have still to be resolved on the stability and integrity of solidified wastes.

The required period of isolation from the biosphere for high-level and alpha wastes has been the subject of some dispute. Some nuclear proponents have claimed that 600 to 900 years is a sufficient period, but this view is based on unsupportable assumptions and is not generally accepted, even within the nuclear industry. Major long-lived constituents of high-level wastes, with their respective half-lives, are the transuranics – neptunium-237 (2.14 million years), plutonium-239 (24 390 years), plutonium-240 (6600 years), plutonium-242 (390 000 years), americium-241 (458 years) and americium-243 (7370 years) and the fission products – technecium-99 (210 000 years), zirconium-93 (900 000 years) and iodine-129 (17 million years). It takes ten half-lives for an isotope to decay to approximately one-thousandth of its original activity and twenty half-lives to decay to one-millionth. The required isolation time for any particular isotope is generally taken as ten to twenty half-lives, with the upper figure being used for isotopes that are initially very plentiful or particularly hazardous.

Evaluating the necessary isolation time for a waste management programme necessitates taking into account not only the quantities of each isotope and its half-life and toxicity, but also the radioactive daughter products of its decay. Taking some of these factors into account has led to plutonium-239 being regarded as the longest lived crucial isotope. This has implied an isolation time for wastes of 250 000 to 500 000 years. However, neptunium-237 with a half-life of over two million years is estimated to be present in the initial waste stream at some fifteen to thirty times the quantity of plutonium-239. Its toxicity, at least for ingestion, is as high as that for plutonium-239. Furthermore, according to Amory Lovins in the recent book *Non-Nuclear Futures*[11], the neptunium-237 decay chain gives rise to nearly five times as much alpha and beta radiation in the first million years as that of plutonium-239 and may be the dominant hazard in high-level wastes. The decay of neptunium-237 to one-thousandth its initial level could take around twenty million years.

Options for the Disposal of Long-Lived Wastes

There are essentially three classes of wastes produced by nuclear power which require isolation from the biosphere for periods of one million years or more: high level wastes, fuel cladding, and plutonium-contaminated solid wastes. For such uniquely toxic substances, their volumes are not insignificant. One 1000-MWe reactor operating for one year could be expected to give rise to some forty-five cubic metres of high-level liquid waste concentrate and some two cubic metres of fuel cladding waste. If plutonium recycling goes ahead, plutonium-contaminated solid waste arising from the 200 kilograms of plutonium produced in one reactor-year could amount to 200 cubic metres at current rates of production. The projected world nuclear capacity of 2000 GWe in the year 2000 would give rise to

60 000 tonnes of spent fuel annually. The Fox Report estimates the world's cumulative production of high-level liquid waste to the year 2000 at over 500 000 cubic metres. Plutonium-contaminated solid wastes are predicted to reach three million cubic metres by then, containing up to twenty-two tonnes of plutonium. High-level waste can be reduced in volume by five to ten times by compression and incineration, depending on the type of materials involved. However, not all materials are amenable to this. The requirement to dispose of these three classes of waste does impose limitations on some of the disposal options that have been suggested. Specifically, those options that require wastes to be packaged in canisters of set sizes and shapes may need to be re-examined.

A fourth class of waste may present more extreme limitations in the future. It is now being seriously suggested that thousands of tonnes of material from the demolition of defunct nuclear facilities be disposed of by the same methods as high-level wastes. Certainly the alternative of leaving old structures on the surface appears quite unsatisfactory.

Extra-Terrestrial Disposal

High on the list of exotic options for waste disposal is that of extra-terrestrial disposal. It has been argued that wastes should be removed from the earth permanently by rocketing them off to that big-reactor-in-the-sky, the sun. In reality, it is only science fiction enthusiasts and the ailing space lobby who still consider this is a serious possibility. While the sun is undoubtedly an excellent place to dispose of the wastes, the requirement of getting them there, safely and cheaply, rules it out. There is no way anyone could guarantee the rocket with its dangerous load, a successful and permanent escape from earth's gravitational field.

Disposal in the Polar Icecaps

The idea of leaving wastes on the Antarctic ice cap to melt their way through to the rock below has also been widely suggested. However, for several reasons this proposal is no more realistic than the previous one. Little is known about the processes at the ice/rock interface and what water movements occur there. It is quite likely that water would flow from the final resting place of the wastes directly to the sea, defeating the whole aim of isolation.[12] In any case, studies by the Battelle North West Laboratories, USA, show that glacial movements of ice towards the sea would carry the wastes into the biosphere within 100 000 years. The Australian government, for one, has strongly opposed this suggestion for disposal, thereby blocking any international agreement necessary for such a move. It seems improbable that international consensus would ever be reached with respect to such a use of the Antarctic.

Transmutation

A more serious suggestion is the proposal to degrade wastes by transmutation. In general, techniques for transmutation involve the production of high-energy particles, which in turn break apart the undesirable nuclei, a process known as 'sub-critical fission'. The aim is to produce wastes that require isolation for a much reduced length of time – i.e., 100 to 1000 years, as opposed to the present 500 000-year to 2 000 000-year hazardous lifetime. To achieve the reduction to 1000 years requires the separation and destruction of the long-lived actinide isotopes, including those of plutonium, neptunium, americium and curium, as well as a number of very long-lived fission products, technecium-99, zirconium-93, iodine-129 and caesium-134. To reduce the necessary storage time to the shorter extent of 100 years requires the additional destruction of large quantities of strontium-90 and caesium-137.

The first major obstacle to transmutation as a waste disposal option is the problem of achieving adequate removal from the waste stream of those elements requiring transmutation. Unless removal is virtually a hundred per cent, then the required isolation time will not be reduced. It takes some twenty half-lives to reduce an isotope to one millionth of its original activity. Separation from the waste stream to similar orders of magnitude would be necessary. This is well beyond the technical capacity of reprocessing by solvent extraction, which removes uranium and plutonium with up to 99.5 per cent efficiency, leaving one two-hundredth of the plutonium in the wastes.

A secondary recovery process, using ion exchange, has been suggested by workers at ERDA's Oak Ridge National Laboratories in the USA. This process is still under development and has many problems, including the production of increased high- and low-level wastes and the lack of adequate processes to remove americium and curium. This processing was expected to give recovery rate percentages of 99.9 for uranium, 99.99 for plutonium, 95 for neptunium, 99.9 for each of americium and curium, and 99.9 for iodine. Even then, the resultant 'short-lived' wastes would have fifty times the relative hazard of uranium ore after a thousand years. ('Relative hazard' here is the quantity of water required to reduce each isotope of the wastes, or the ore, to the maximum permissible concentration (MPC) for that isotope.) The basic precondition for the success of transmutation does not therefore exist at present. Further requirements for transmutation are that the process must not use excessive energy, nor should it substantially increase the risks to workers or the general public compared to alternative means of waste disposal.

The most seriously considered transmutation option is to recycle the actinides through nuclear reactors. While in theory this is quite feasible, the idea is still at the basic research stage. The proposal to use commercial reactors does put severe limitations on this concept. At best, commercial power reactors could be expected to burn up only 10 per cent of the actinides placed in it. The wastes would have to be recycled up to ten times before sufficient quantities would be transmuted. This means up to ten times through the fuel cycle, reprocessing, transport, fabrication, fission, etc., with all the associated risks of exposure and the production of wastes. It has been suggested that actinide recycle would only be worthwhile if burn-up rates of 90 per cent rather than 10 per cent could be achieved, in order to eliminate this multiple handling. In the meantime, if the recycling of plutonium is rejected, the whole transmutation question becomes essentially academic, as all plutonium would then become waste, requiring disposal, whereas only approximately 1.5 per cent is presently considered so.

Seabed Disposal

One proposal for waste disposal in the sea envisages the placement of wastes into certain deep trenches in the ocean floor which mark the junction of two converging tectonic plates, to be eventually carried down into the earth. Responsible studies have rejected this proposal because of the extreme paucity of information on the mechanisms behind this process. If the mechanism were to fail, slow, deep-water currents could carry the wastes to the surface within 100 to 10 000 years.

A second and more realistic proposal is to drill shafts into stable areas of the ocean floor, stack canisters of waste in the shafts and, when full, plug up the top of the hole. Certain plates in the major oceans have been stable for long periods and have been the subject of some research, By comparison, however, the oldest rocks discovered on the ocean floor are only 150 million years old compared with the age of the continents which are three to four thousand million years old. Safety problems exist both in transport and emplacement of the wastes in the seabed; furthermore, the wastes would be virtually irretrievable.

The Battelle Institute's North West Laboratories, in a major report for the USAEC in 1974, called 'Advanced Waste Management Studies', rejected the option of seabed disposal because of the lack of natural barriers between the wastes and the biosphere. In this case, the burial container and the plug are the main barriers, both of which were considered to be potentially unreliable over the required period of a million years or more. On the other hand, the recent Flowers Report on nuclear power in the UK recommended serious attention be given to investigating seabed disposal, as well as land-based options. In 1974 two US researchers, concluding an extensive introductory study of this concept, reported[13]:

While far from advocating seabed disposal, and almost as far from a conclusion that it is technically feasible, we have identified a type of area on the planet that we feel deserves careful study as a possible respository . . .

Geologic Disposal – Salt

Much research has gone into the option of disposal of wastes into the ground. While many types of geological formation have been suggested, such as granite, limestone, dolomite, shale, gneiss, schist and mudstone, it is salt formations and especially bedded salt which are considered to have the most potential. Salt is usually pure to greater than 99 per cent; it is plastic, and hence not susceptible to most geological disturbances; it does not interact significantly with the biosphere; it is a good conductor of heat and, most importantly, salt formations are usually dry, generally having no contact with ground water.

Three most important criteria for a geological disposal site are: absence of water, structural stability and a low resource potential for the area. Circulating ground water represents the greatest danger to the safety of wastes buried in salt; because it has the capacity to dissolve substantial portions of the salt formation and carry the wastes up into the biosphere. While ground water is regarded as the major mechanism for transport to the surface of wastes placed in any geological formation, salt, being water-soluble, is uniquely threatened by such a process.

It is often assumed that salt formations which, by virtue of their existence, have been isolated from water for a hundred million years, will remain isolated for the next million years or more. Such an assumption, however, is far from justified. According to George De Buchananne, nuclear wastes expert of the US Geological Survey, complete assurance of the immobility of wastes is only possible if geologists accept what is called the 'uniformitarian principle' – that is, that all geological changes in the past are accountable in terms of processes of the same type and intensity as those observable at present. However, one of geology's puzzles concerns the formation of oil and shale deposits usually associated with underground salt formations. These deposits defy explanation in terms of the uniformitarian principle. Furthermore, man's activities may well be capable of negating this principle by upsetting, to an unknown extent, the geological stability of the earth's crust. Research workers at ERDA's Oak Ridge National Laboratories in Tennessee reported[14], in 1974, that they had correlated minor but significant earthquake activity, in a previously seismically stable area, with water injection operations at local oil wells. Likewise, underground dumping of toxic chemical wastes in deep wells has had similar effects. For example, 710 earthquakes recorded in Denver, Colorado, over five years have been linked to concurrent disposal operations at the nearby Rocky Mountains Arsenal. Serious proposals in the USA to explode nuclear bombs in natural gas wells to increase the gas yield could have significant effects on the geological stability of a wide area around.

Human activity impinges in other ways on the safety of wastes buried underground. As previously observed, fossil fuel reserves of oil, gas and carboniferous shale are often associated with salt formations; conversely, it is practically impossible to find a salt formation where the potential for finding oil and gas cannot be totally precluded. Potash, another valuable resource, is also associated with underground salt. These factors, as well as the inherent value of the virtually pure salt itself, makes these formations subject to a great deal of attention by mineral prospectors both now and over the hazardous lifetime of the wastes buried there. Such drilling would be a substantial threat to the isolation of any wastes buried there. To be potentially dangerous, bore holes would not necessarily have to intersect the wastes, but merely provide a pathway for groundwater to them. Studies have shown that, of 50 000 wells drilled through salt in central Kansas, in five known cases a gravity-driven, circulating fresh-water system has rapidly dissolved the salt, leading to large depressions in the surface terrain. These are hazards that cannot be eliminated. They exist because there is no way in which man could reliably pass on records of his disposal operations, or otherwise prevent an accidental encounter with such wastes. There is also the problem of obtaining sufficient information in the process. The only reliable geologic information is obtained by drilling in, around and through the site, thereby potentially providing pathways for water into the formation. At some point a trade-off must be made between increasing knowledge of and increasing damage to the integrity of the formation.

Disposal in underground salt must be considered as a virtually non-retrievable operation. Not only would the waste's steel containment vessel be expected to corrode within a couple of years, but McClaim and Boch have shown that the mining level would become uninhabitable within twenty to thirty years due to decay heat and would remain so for several centuries.

While disposal in salt is being seriously considered in the USA and West Germany, it is not equally relevant to many other countries with nuclear power programmes. Each of these countries has its own set of constraints to work within and most have their own views on what options may be most suitable. Japan and most western European countries have much more severe constraints in terms of land area, range of geological circumstance and population density. Japan is particularly unsuitable because of seismic instability, and some European countries have recognized that potential geological sites are so scarce that they could only cope in this way with particularly difficult wastes, if at all.

Other Geologic Disposal Methods

The USAEC's 1974 report[15] on high-level waste management alternatives, considered nine proposals for geologic disposal other than the one above. Three of these involve the sealing of the wastes underground to enable the decay heat to melt the surrounding rock. Rock melting would be expected to occur over a few decades to a hundred years, when mixing of the waste and rock would take place, the whole melt solidifying as a homogeneous matrix after a few hundred years. Not only would these options hand on to future generations a potentially explosive problem of hot liquid wastes but they offer no safeguards of retrievability at all. Further proposals examine the dumping of wastes in shafts from three to sixteen kilometres deep. Very poor retrievability characterizes all these concepts and, in addition, there are many unknowns in present understanding of deep geological processes and formations. The study also examines the placement of solid waste in man-made structures in geological formations. These proposals would require a positive commitment by society to maintain cooling and containment for many centuries into the future. Most of the concepts studied were estimated to require fifteen to thirty years for full implementation, if found to be technically feasible.

Morality and the Future

The handling of radioactive waste involves a great variety of materials of widely varying hazard to the environment. The uniqueness of this hazard is that environmental and health effects are generally transferred decades or even generations into the future, either because of the long life of the waste or because cancers and genetic mutations produced take time to appear. It is the present generation that will receive the questionable benefits of nuclear power. But by creating nuclear waste in all its forms, we are placing a series of burdens and responsibilities on all future societies, such as the maintaining of records of disposal sites and the alienation of large areas of land from all future uses, to ensure isolation of the wastes.

The problem of disposing of wastes is one of the most serious to be contemplated by western society in any commitment to nuclear power. Australia, as a past and present supplier of uranium is morally implicated in the nuclear wastes issue. According to a 1976 report[16] from the South Australian Uranium Enrichment Committee, every

1000 tonnes of uranium mined and milled will become some 150 tonnes of enriched uranium which will fuel a 1000-MWe LWR for around five years, and 850 tonnes of depleted uranium to be converted to plutonium if breeder reactors go ahead. The spent fuel, if reprocessed, would give rise to 210 cubic metres of high-level liquid waste and over one tonne of plutonium, as well as copious quantities of low-level wastes at all stages of operation. The depleted uranium, when and if converted to plutonium, could give rise to some 700 000 cubic metres of plutonium-contaminated solid waste. Our potential to provide the material for these wastes gives us a responsibility to the rest of the world. The First Fox Report concludes that:

The creation anywhere of huge amounts of long-lived radioactivity is a matter of concern to all people, because of the very serious global effects which dissemination of them could have. (p. 178)

It is becoming evident that there has been a recent change of attitudes to the nuclear industry's glib statements that solutions to major problems of nuclear waste will be forthcoming. The UK Flowers Report concluded:

There should be no commitment to a large programme of nuclear fission power until it has been demonstrated beyond reasonable doubt that a method exists to ensure the safe containment of long-lived, highly radioactive waste for the indefinite future. (p. 202)

In August 1976 the US Court of Appeals declared that the Nuclear Regulatory Commission was remiss and negligent in failing to consider the environmental effects of radioactive wastes in its decisions on licensing of nuclear reactors. In October 1976, in Australia, the Fox Report concluded that:

If, even in a few years, satisfactory disposal methods have not been established, it may well be that supplies of uranium by Australia should be restricted or even terminated. (p. 178)

Can we trust the nuclear industry to find satisfactory solutions to the many problems of wastes in the fuel cycle? Can we trust an industry which, for example, allows its depleted uranium to be used in the porcelain dentures of fifty million Americans, or which allows radioactive uranium mill tailings to be used as landfill under tens of thousands of houses? These are not isolated instances, but are characteristic of the way in which nuclear power has developed. However optimistic we may be about the industry's good intentions regarding waste management, we must take into account the record of the last thirty years.

Practical demonstrations of long-lived waste disposal options are still two or three decades or one generation away on present estimates, with no guarantee that they will prove satisfactory even then. By that time, however, the cumulative wastes requiring disposal will be many times greater than at present. Nuclear power, therefore, while posed as a solution to energy problems, represents in reality a relocation of present problems into the future. Its wastes will burden the earth and its future inhabitants for longer than *Homo sapiens* has been known to exist.

5 Bombs: Nuclear Proliferation and Nuclear Theft
Greg Woods
Mary Elliott

Sir Philip Baxter has said before the Ranger Inquiry, 'We have to learn to live with nuclear weapons. I see no great problem about that.' We have, in a sense, learned to live with nuclear weapons. In Europe, after the popular protest of the 1950s and 1960s failed to prevent Britain and France joining Russia and the USA in developing nuclear weapons, people resigned themselves to the constant threat of nuclear war, feeling powerless to do anything about it.

Once the 1963 Test Ban Treaty stopped atmospheric tests by Britain, Russia and America, the whole subject went underground. Nuclear weapons were regarded as a deterrent to war and, in this respect, even thought beneficial. The superpowers were stalemated by the devastating nuclear firepower they possessed which, it was hoped, none would dare use. The prospect of nuclear war became familiar but unreal to all but the military on both sides, who devised more and more sophisticated weapons and delivery systems, at a cost of billions of dollars, in order to gain the supremacy of the pre-emptive strike. The result now is that nuclear-armed ships, planes and submarines of both Russia and America patrol the world at hair-trigger readiness. The Hiroshima bomb is described as a 'baby' compared with the fifty and hundred megaton monsters trained on cities. NATO forces possess small tactical nuclear weapons designed to knock out tanks. They may soon be equipped with the neutron bomb, which kills people by intense radiation without destroying property.

Although the human race is inured to death, disease and constant war, it has never before this generation faced the prospect of total extinction. Amid the worst horrors of the last two world wars, people struggled to survive, knowing there could be survivors to rebuild the ruins when the guns stopped. Nuclear war is different. The strategists discuss where and how a remnant of the human race would survive the nuclear holocaust, yet nuclear war is unthinkable: its effects would

Figure 5.14 Effect of a 50-megatonne bomb on Sydney compared to a Hiroshima-size bomb

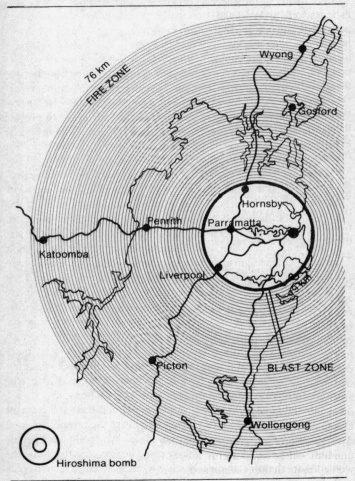

Source: Computed and drawn by Mike Bell and Rod Simpson

not only be a tortured death for millions of people, but continuing genetic damage to any survivors for all generations.

It is a matter of the greatest concern that such obvious facts need to be restated, that we must continually remind ourselves of the desperate position we are in. However, there are signs that there is a new awareness of the danger of nuclear war, which may lead to positive action. The vague though ever-present threat of nuclear destruction has not been enough over the last quarter-century to prevent the growth of huge nuclear arsenals. But, as nation after nation, including Australia with its rich reserves of uranium, the fuel for nuclear bombs, is drawn into the nuclear world, the reality of the nuclear threat will become apparent. In Australia, for example, the controversy about the export of uranium has made it no longer possible for Australians to observe passively the play of the great powers from a distance. With the export of uranium, Australia would be positively involved in the nuclear programmes, civil and military, of the USA, Western Europe and Japan. That fact in itself must serve to heighten the consciousness of Australians to the danger of war. The uranium issue brings home, in an unprecedented way, that no nation can escape the most urgent and terrifying of the world's problems, the immediate danger of nuclear war.

It is essential that this possibility is taken seriously before anything can be done. The danger is, on the other hand, that the problem will appear so overwhelming that we flinch from tackling it, living only for today since there may be no tomorrow. This kind of malaise is much in evidence in the western world. Although the deep effect of the nuclear threat is not measurable, it is undoubtedly there and has been since August 1945. History has taken on a new and more desperate dimension. It is very difficult to plan cogently for a future that may not exist. The result may be that people do not plan at all and resort instead to a blind faith in experts and governments to lead them out of the mess which is beyond their control. Australian uranium has in the past, and could again, fuel the bombs. Yet many Australians prefer to accept comforting government assurances that uranium will be for 'peaceful' uses only, that 'safeguards' on its use work, despite all the evidence to the contrary.

Bad though the current nuclear weapon situation is, there is worse around the corner. Nuclear bombs are a byproduct of nuclear power.

From the reprocessed reactor fuel is obtained plutonium, which can be made into weapons. The more reactors producing plutonium there are in the world, the greater the possibility of an increase in the number of bombs. The more countries with nuclear power, the more countries there will be that could possess nuclear weapons.

It is a great responsibility to make sure that uranium does not get into the wrong hands. There is an elaborate system of 'safeguards' administered by the International Atomic Energy Agency (IAEA) and a treaty, the Nuclear Non-Proliferation Treaty (NPT), which makes parties to it undertake not to make bombs from reactor fuel. But whether or not the safeguards do and will work is a matter of great controversy. People involved in devising and administering safeguards are intensely worried. In a new book called *The Last Chance*, by William Epstein, the special consultant on disarmament to the Secretary-General of the United Nations, the first words are:

For the first time in a quarter of a century of working with the problems of the arms race and arms control, I am beginning to get scared ... there is a clear and present danger of the entire structure of the so-called non-proliferation regime – the structure of treaties and agreements, crumbling.

At the NPT Review Conference held at Geneva in May 1975, the major five-yearly reappraisal of how the Treaty is working, no significant advances were achieved. In fact the conference almost broke down and was saved only at the last moment by a vague declaration of good intentions by Sweden, which became the conference's public statement. The heads of the Russian, American and British delegations showed their indifference to the proceedings by leaving only a few days after the conference had begun. Thirty-seven states who had ratified the Treaty, mostly Third World countries, did not even attend the conference. Only fifty-eight of ninety-five parties were present. Even they were not continuously represented at the meetings because their delegates felt other business took priority. Such is the importance that the nations of the world place on developing safeguards on nuclear energy and working for nuclear disarmament.

It is not surprising then that at the fifteenth General Conference of the IAEA in Rio de Janiero in September 1976, the Director-General himself said, in his opening address, 'In my opinion we should recognize that we may be nearing the end of the NPT line.' But this view is not universal. For example, when the Australian Minister for Environment, Kevin Newman, announced on 11 November 1976 that Australia would honour existing conrtacts for 9000 tonnes of uranium, he said: 'The government is satisfied that appropriate controls would apply to the shipments under existing contracts which will be used for electrical power generation in Japan, the United States and the Federal Republic of Germany, all of which are parties to the Nuclear Non-Proliferation Treaty.' This was followed by the return in December 1976 of an Australian government mission that had visited Britain, Europe, the USA, Canada and Japan to investigate nuclear safeguards, to pronounce itself confident that nuclear materials and technology could not fall into the wrong hands. The same confidence was shown by the Australian Atomic Energy Commission (AAEC) at the Ranger Inquiry. In their final submission to the Inquiry, the AAEC averred that 'current indications are encouraging', although admitting in a monumental understatement that 'it would be imprudent to assume that no problems remain.'

The Nuclear Non-Proliferation Treaty

The atomic bombing of Hiroshima and Nagasaki in August 1945 was immediately recognized as a turning point in human history. Albert Einstein said, 'The splitting of the atom has changed everything save our modes of thinking and thus we drift toward unparalleled catastrophe.' Thinking certainly had not changed and there was immediate talk of controlling the destructive potential of the split atom and of developing nuclear energy for commercial use. Although leading scientists like Einstein gave clear warning that human institutions and human beings with all their imperfections were totally inadequate to cope with a nuclear world, work on developing nuclear energy went ahead. Delegations from Canada, the UK and the USA met in November 1945 to discuss control of weapons, but also the development of commercial nuclear energy. In the thirty years since then the same pattern has been followed – an international attempt to devise controls on the military use of the atom while at the same time encouraging commercial use. The failure of these controls is obvious. They have not stopped the superpowers from developing huge arsenals of ever more sophisticated and varied nuclear weapons. They have not prevented Britain, France, China and in May 1974, India from developing nuclear weapons.

Bertrand Russell, in the early 1960s, warned that, unless Britain pursued a policy of unilateral disarmament, the arms race would continue unabated and eventually nuclear weapons would spread to many nations throughout the world. Despite strong support for this view within the British Labour Party, its leaders, Hugh Gaitskell and Aneurin Bevan, argued that they 'could not go into the council chambers of the world naked', and that Britain must possess the bomb so that it could be represented at disarmament conferences and force the USSR and the USA to disarm. As a result Britain's defence expenditure has helped to beggar her, she has not influenced the superpowers one whit and, as Russell predicted, nuclear weapons have proliferated. What Einstein and Russell both realized was that, without absolute condemnation of nuclear weapons and a refusal to spread the means to make them throughout the world, all the controls, safeguards and laws devised by man would be meaningless in a world torn by national, racial and economic strife.

The NPT is the current attempt to control the uncontrollable. After several years of complicated negotiations, described in detail in William Epstein's book, the United Nations Disarmament Committee presented the draft treaty to the General Assembly in March 1968. It was approved and finally came into force on 5 March 1970. Its membership has slowly grown and to date 106 nations have signed or ratified it. Australia ratified the Treaty on 23 January 1973. (See Appendix B for text of the Treaty.)

The Treaty was the subject of much debate and investigation at the Fox Inquiry. The First Report provides an excellent commentary on it and on the operation of IAEA safeguards. As a result of evidence presented on the weakness of the Treaty and of safeguards, the Report concluded in its principal findings and recommendations that 'The nuclear power industry is unintentionally contributing to an increased risk of nuclear war.' Not satisfied with current safeguard measures, the Commissioners recommended that 'Nuclear materials should be supplied to a state only on the basis that its entire nuclear industry is subject to back-up safeguards that cannot be terminated by unilateral withdrawal' (p. 148). The Report provides a critique of the NPT system which is of the utmost importance, for 'Existing safeguards may provide only an illusion of protection' (p. 147).

One of the main reasons why the NPT and the IAEA safeguards are so limited is that the Treaty and the Agency are both attempting to put into practice two aims that are in conflict. The first is to encourage the development, worldwide, of nuclear energy, to promote the free flow of nuclear materials, nuclear plants and nuclear technology. Article IV of the NPT commits all parties to:

Cooperation in contributing alone or together with other states or international organizations to the further development of the applications of nuclear energy for peaceful purposes, especially in the territory of non-nuclear-weapon states party to the Treaty, with due consideration for the needs of developing areas of the world.

The reason for this approach was a belief that, if nations were given free access to peaceful nuclear energy, they would be less likely to develop, in resentment at the energy and military superiority of the great powers, not only commercial plants but also nuclear weapons. In other words the Treaty was based on a policy of nuclear appeasement.

Its second aim is to prevent the spread of nuclear weapons and to promote general nuclear disarmament. In practice it is extremely difficult, most likely impossible, both to allow a free flow of nuclear material and information and to prevent effectively a proliferation of weapons. The inevitable result of this conflict in aims is that the safeguards are designed to interfere as little as possible with the operations of the nuclear industry, thereby leaving dangerous loopholes that make the safeguards ineffective.

The Treaty itself has major failings. First, it is not universal: certain key nations are not members. Yet the Treaty does not prevent the sale of nuclear material and equipment between parties and non-parties. If the NPT were a closed shop, with only those agreeing to its terms permitted access to nuclear energy, the Treaty might be more viable. Its existing situation, however, is comparable to the League of Nations, which proved ineffective because a number of nations, including the USA, were not members. Therefore, when the League attempted to impose economic sanctions on Italy after its invasion of Abyssinia, they were a miserable failure. Similarly, as long as there are nations who will not subscribe to the NPT, but who are willing to supply all things nuclear under whatever conditions, or lack of them, they themselves choose, then NPT safeguards will not prevent proliferation.

The key nations outside the NPT system are China, India, Pakistan, Egypt, Israel, Brazil, Argentina, France, Spain and South Africa. China exploded its first nuclear device in 1964, and since the death of Mao Tse Tung the government has made increasingly loud noises about China's ability to become a major nuclear-armed power. Since India set off an explosive device in 1974, thus proving that a poor Third World country could muster the technology to do so, and that reactor equipment and nuclear materials supplied from abroad (Canada and the USA) could easily be diverted to military purposes, Pakistan has sought from the French a reprocessing plant. This has aroused fears that Pakistan's real purpose is to extract plutonium from its nuclear wastes in order to match the Indian example and get a bomb. Israel is reported to possess a number of nuclear weapons, but refuses to deny or confirm this. Egypt refuses to become a party to the NPT because Israel will not. Brazil and Argentina are both acquiring the complete nuclear fuel cycle. Argentina is listed by William Epstein as a nation that could develop nuclear weapons within one or two years. In the same category is South Africa, which could turn to nuclear weapons as a last resort to defend the white regime against the black nations which surround it.

France has possibly done more than any other country to sabotage the NPT system. Despite international protest, the French tested weapons in the atmosphere over the Pacific in the early 1970s. France has pursued an active nuclear power sales campaign in the Middle East, South Africa, Pakistan and South Korea and adamantly refused, despite, or because of, pressure from the USA, to join the NPT. In 1976 US pressure on South Korea prevented the sale to that country of a French reprocessing plant. At the time of writing the French deal to supply Pakistan with a reprocessing plant still stands. The irresponsible attitude of France toward nuclear marketing and its large domestic nuclear power programme have aroused worldwide concern and active protest by the French people. As a result France may scale down its ambitions for a new, atomic empire. But it has provided an example of what one nation, possessing the materials, equipment and technology, can do to make the safeguards terms of the NPT a meaningless exercise in idealism. Yet France can argue that it is merely implementing Article IV of the NPT, promoting the free exchange of nuclear goods for the 'benefit' of all. NPT members are therefore in a weak position to complain of French nuclear adventures.

Because this uncontrolled situation exists, the First Fox Report expressed concern that Australian uranium could be re-exported by a country such as Japan to a nation that Australia might not approve of. If it were in the form of yellowcake, no safeguards would apply if the material were re-exported to an NPT member. To ensure absolutely that Australian uranium is fully accounted for, the Report recommended that Australia should insist that no re-transfer at all should take place without Australian approval of the eventual purchaser. The difficulty lies in enforcing such a provision once the uranium is out of the country. The Report warns that:

> Although it would be possible to include provisions regulating retransfer in the commercial contract relating to the original sale, it would be unsatisfactory to rely upon the efficacy of such provisions alone, because there could well be problems in seeking to enforce them in the courts of foreign states. (p. 131)

This highlights the problem of safeguarding nuclear materials. Agreements and contracts can always be broken and once uranium leaves the national boundaries of Australia, its ultimate fate is reliant upon expediently made promises.

The second major weakness of the Treaty is that no sanctions against a party that disobeys the rules are written into it. Should this happen, should a member be found to be diverting nuclear material to make bombs, the IAEA could withdraw its assistance and call for the return of nuclear materials and equipment. Should there be a consensus in the UN Security Council (an unlikely prospect), the UN could call upon its members to enforce economic sanctions. However, the history of the UN and other international attempts at economic blockade is one of failure. Rhodesia survived UDI despite such measures. Uranium is still being exported from Namibia (South West Africa) despite a UN ban on all exports because of South Africa's illegal presence there. This, together with the fact that the NPT does not attempt to spell out ways to enforce its terms, means that there is no effective way its provisions can be enforced. It is merely an expression of good intent. It does not have the power of a law within a nation state. There is no way in which any nation or group of nations can take back the reactors, nuclear fuel, technical knowledge and nuclear weapons of another country, disarm it as the police would an armed robber, and put it into gaol. Individual suppliers like Australia could, it is true, refuse further supply and this point comprised the sixth finding of the First Fox Report.

> A decision to mine and sell uranium should not be made unless the Commonwealth Government ensures that the Commonwealth can at any time . . . immediately terminate those activities, permanently, indefinitely or for a specified period. (p. 185)

This is obviously the only sanction that can be applied by a supplier state that genuinely wants to ensure its uranium does not become bomb fuel. However, not only would this be merely to shut the door after the horse had bolted, it is also highly unlikely that countries committed to nuclear power would accept such terms since an assured supply of uranium would be essential to them. Nor would mining companies invest large amounts of capital in a project that could be stopped at any time. The only realistic solution, therefore, is not to export any uranium at all until enforcement of safeguards is an assured reality.

The third weakness of the NPT lies in its reliance on Article VI which requires the parties to work for a treaty on general and complete nuclear disarmament 'at an early date'. All comment on the Treaty agrees that if this clause is not put into practice, then nations with nuclear power cannot be restrained from building their own bombs. Out of national pride and fear for national security, as well as a desire for independence from the dominance of one superpower or the other, nations may choose, in the absence of general disarmament, to become their own nuclear bosses. This is obviously the basis of French policy, a country that sees the alternative as hiding under the nuclear skirts of the USA and NATO. There is a small but vocal lobby in Australia which believes that total reliance should not be placed on the USA, that the only way is for Australia, with its long coastline and undefended wildernesses, to invest in the 'security' of its own nuclear weapons. There is a similar, though larger and more powerful, nuclear lobby in Japan.

Throughout the world the trend is the same. As long as the superpowers refuse to disarm and there is no binding and perpetual agreement, as was sought in 1945, to ban all nuclear weapons on earth,

Bombs: Nuclear Proliferation and Nuclear Theft 137

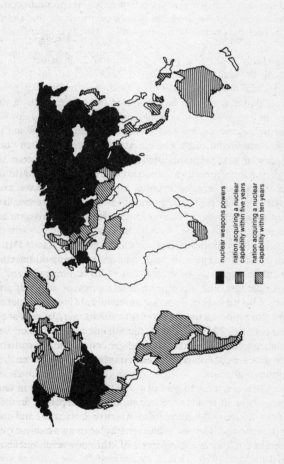

Figure 5.15 Present and potential nuclear weapons powers

- nuclear weapons powers
- nation acquiring a nuclear capability within five years
- nation acquiring a nuclear capability within ten years

Source: Compiled by Mike Bell and Rod Simpson from estimates by William Epstein in *The Last Chance*, The Free Press, New York, 1976, and International Atomic Energy Authority reports.

there is the imminent danger that all nations able to do so will develop weapons. According to Epstein, the number of nuclear nations that could do so within ten years is thirty-nine. His prediction is based on the ownership by these nations of commercial nuclear plants, nuclear fuel and nuclear technology, all originally donated for 'peaceful' purposes. It is almost inconceivable how any government can express faith in the NPT when the general disarmament clause, on which all the rest depends, has been broken continuously from the day on which it was written. This is a major reason why the Director-General of the IAEA said last year that the NPT may be at the end of the line.

The most serious weakness of NPT is the ease with which parties can withdraw from membership. Article X states that any party to the Treaty, if its supreme national interests are affected, can withdraw after giving three months notice to the UN Security Council and other parties. On doing so, all safeguards, all controls on a nation's nuclear activities, would cease. The point has been made that the Treaty is merely a promissory agreement, but even the moral value of such a commitment is weakened by Article X. In political terms this means that, should, for example, Japan consider itself threatened by China and in dire national peril, it could turn its commercial nuclear establishment immediately to military production, arguing, within the terms of the NPT, that supreme national interest necessitated its withdrawal. The First Fox Report commented, 'This is undoubtedly a serious limitation on the operation of the NPT and of most safeguards agreements' (p. 128). It is why the Report recommended that additional agreements which could not be broken by the unilateral act of one state be concluded between Australia and countries to which it supplies uranium.

Because it is so easy to get out of the NPT and its obligations, many nations whose intentions are, to say the least, dubious, have joined it. Seven nations, Belgium, West Germany, Ghana, Italy, Libya, Luxembourg and Rwanda ratified the Treaty at or before the Review Conference in May 1975. In March 1976 Singapore ratified and Japan ratified in May 1976. This is interpreted by the AAEC as a sign of the growing strength of the Treaty. Unfortunately, there is some evidence that the motives of some recent signatories and the permanency of their adherence to the Treaty are subject to doubt. For example,

Bombs: Nuclear Proliferation and Nuclear Theft 139

Libya ratified in May 1975, but in the previous January President Gaddafi was reported in the press as saying,

I imagine that nuclear power will one day be as essential as electricity, and while people now say this country has fifty planes and that country 500 planes, the day will come when they say that this country has three nuclear bombs and that country ten and so on.

As a party to the NPT, Libya will receive full cooperation in its development of a nuclear capacity, help which it needs at the moment. Having built up its nuclear industry, Libya could then plead national interest, withdraw from the Treaty and build bombs. It is possible that, in this way, the Treaty could be used as a means for nations to acquire nuclear power, with the long-term aim of becoming independent nuclear-armed powers. In the case of Japan, whose ratification of the Treaty has become the main plank of the Australian government's argument that uranium export there is safe, the interval between signing and ratification was six years, the Treaty was ratified only after numerous abortive attempts and immediately prior to the visit to Japan of the Australian Prime Minister to discuss uranium exports. The opposition to the Treaty in the Japanese Diet remains and an increase in power to the Japanese right wing could result in Japanese secession from the NPT. This is a vivid illustration of the tenuous nature of the Treaty and the use to which it is being put by nations anxious to acquire uranium. The formalities are being observed; whether the spirit of the Treaty will be equally respected over the next crucial decade is open to grave doubt.

The NPT is, then, a mere promise to behave. If 'national interest' is felt to be at stake, any country can leave the NPT system, in possession of nuclear power plants, reprocessing plants, stockpiles of uranium and plutonium. There is in effect nothing a supplier nation like Australia could do about such a situation, once the uranium is out of the ground and out of the country. Yet it is on the fragile base of NPT promises that Australian governments have assured the people that Australian uranium cannot be used for bombs. Such a claim is almost too absurd to be taken seriously, were it not that some Australians may believe it, and may be led, for the supposed sake of immediate economic benefit, to support a 1970s version of the 'pig iron policy' of the 1930s.

Table 5.3

States—members of the United Nations—not party to NPT, as at 31 August 1976

Albania Algeria Argentina Bahrain Bangladesh
Barbados Bhutan Brazil Byelorussian SSR Cape Verde
Chile China Colombia Comoros Congo Cuba
Egypt Equatorial Guinea France Guinea Guyana
India Indonesia Israel Kuwait Malawi Mauritania
Mozambique Niger Oman Pakistan Panama
Portugal Qatar Sao Tome and Principe Saudi Arabia
South Africa Spain Sri Lanka Trinidad and Tobago
Turkey Uganda Ukrainian SSR United Arab Emirates
Tanzania Yemen Zambia

IAEA Safeguards

The whole nuclear debate is bedevilled by euphemistic use of the word 'safeguards'. It is usually qualified in government statements by a number of adjectives, including 'adequate', 'strong', 'strict', 'improved' and 'completely satisfactory'. Because the subject is shrouded in secrecy for reasons of security and propaganda, few Australians know what the safeguards are or how they work. For its description and critique of them, therefore, the First Fox Report is most valuable.

Experience with IAEA safeguards demonstrates that countries have not been prepared to accept continuous surveillance of nuclear activities by an external authority. The control system involves accounting measures augmented by regular 'on the spot' inspections. The inspections are carried out by a team of skilled personnel within the IAEA. As at 30 June 1976, according to information available to the Commission, there were seventy-nine inspectors, of whom about fifty regularly carried out inspections. Several witnesses were of the opinion that this force was far too small to maintain effective surveillance of existing installations covered by IAEA safeguards.

In brief, the accounting procedure is based on a system of records and reports which are maintained by a country with respect to facilities and nuclear materials in its territory. It is then for the inspectorate to carry out regular audits of records and reports, to check the amount of safeguarded material and to scrutinize the operation of facilities subject to safeguards.

Before commencing, the inspectorate carries out a facility design review to ensure that effective safeguards can be applied to each facility. *Precise details of the safeguards measures to be applied in a particular state are included in a subsidiary arrangement to the safeguards agreement, which is not made public, so as to protect any commercial and industrial secrets of state* [Our italics]. (p. 120)

The job of the IAEA inspectors is to visit nuclear power plants in countries around the world periodically and measure the actual amount of nuclear material going through the fuel cycle against the country's records, to make sure that none is missing. This is technically and politically an enormously difficult task and will become more so if nuclear power develops on a massive scale and unless the funds and number of inspectors grows in proportion.

There are many problems. It is in practice very difficult to check the exact amount of nuclear material going through the complexities of the nuclear fuel cycle. Many discrepancies, both plus and minus, have been recorded because, to put it at its simplest, the stuff gets stuck in the pipes. When the system is periodically cleaned out, nuclear material may reappear in the inventories in unexpected quantities. The point is that in the interim there is an amount of nuclear material which is not accounted for. The subject was a matter of dispute at the Fox Inquiry, with the AAEC claiming that the MUF factor (Material Unaccounted For) was not significant. However, Sir Philip Baxter said:

First of all I do not believe that in a country with a large nuclear power industry, and adopting the methods which the IAEA have decided to adopt in the system of inventory control, that it will be possible to balance the inventory of plutonium in a way which will literally account for everything. And the amounts which could appear to be unaccounted for, unaccounted because of difficulties with analytical control and measurement, essentially could be sufficient in theory for someone to make a weapon and it will never be possible to be quite certain whether the material unaccounted for is real or not. This is a defect in technology which, up to now, we have no means of solving. (Fox Inquiry transcript, p. 2104)

The First Report agreed that:

Given present levels of measurement error, it is difficult to believe that regular diversion of a very small amount (say 0.5 per cent) would result in bias in the MUF sufficient to be detected by accounting procedures. In a country with a large nuclear industry ... this could amount to a considerable quantity. (p. 134)

There may already be a considerable amount of plutonium lost, stolen or strayed in the USA. Allegations have been made there that companies may be secretly stockpiling plutonium in order to sell it in the future on a black market where plutonium will be worth ten times more than gold. In September 1976 the US General Accounting Office (GAO) reported that 'There is no clear way that the government can determine whether nuclear material is lost or stolen' and estimated that nearly 500 kilograms of enriched uranium and plutonium were missing, possibly but not certainly through the vagaries of the nuclear industry. So much for the guarantees of accounting procedures.

Because, like the NPT, IAEA rules are aimed at minimal interference in the development of nuclear power, they do not cover all nuclear materials or all stages of the fuel cycle. Source materials – in other words non-enriched uranium and enriched uranium used for 'non-explosive military uses', for example in the reactors of nuclear warships – are exempt. Unless safeguards become universal on all nuclear material from the moment that it is mined to the time it is buried as waste, which would be in effect for all time, the opportunity remains for diversion into weapons by nations and terrorists. Add to these limitations the prospect of uranium and plutonium being transported daily on the traffic lanes of the world by train, ship and plane and the problem of preventing theft can be appreciated. It has therefore been suggested by the IAEA that, instead of transporting nuclear materials between nuclear power plants and distant reprocessing plants, 'nuclear parks' should be established containing the whole fuel cycle, which would be under international control. The nations of the region would be supplied with electricity through an enormous grid. Physical security would be much easier to enforce. But the idea is politically unrealistic. National plans for nuclear development of whole fuel cycles are already well advanced and it is highly unlikely, in any case, that nations would happily cooperate in building a nuclear park under United Nations control. In Australia, it is equally unlikely that a plan to build a squad of reactors on the uranium fields of the

Northern Territory, together with an enrichment and reprocessing plant and high-level waste disposal area, in order to provide electricity for South-East Asia, would gain any political party many votes. Since, according to this idea, the region would become in effect United Nations territory guarded by a multi-national UN force of security men and IAEA inspectors, all to be paid for by massively increased contributions of IAEA members, including Australia, it would appear politically realistic to forget the idea. The spread of nuclear power means continuous transport of dangerous nuclear materials, the high cost of security measures, the ever present danger of theft. Current safeguards and the slight prospects for their improvement in the future cannot eliminate these dangers.

Possibly the most important paragraph of the First Fox Report reads:

The main limitations of the present safeguards arrangements can be summarised as follows: the inability of safeguards to prevent the transfer of nuclear technology from nuclear power production to the acquisition of nuclear weapons competence; the fact that many nuclear facilities are covered by no safeguards; the existence of a number of loopholes in safeguards agreements regarding their application to peaceful nuclear explosions, to materials intended for non-explosive military uses and to the re-transfer of materials to a third state; the absence, in practice, of safeguards for source materials; the practical problems of maintaining effective checks on nuclear inventories; the ease with which states can withdraw from the NPT and from most non-NPT safeguards agreements; deficiencies in accounting and warning procedures; and the absence of reliable sanctions to deter diversion of safeguarded material.

The passage continues:

The Commission recognises that these defects, taken together, are so serious that existing safeguards may provide only an illusion of protection. However, we do not conclude that they render valueless the concept of international safeguards. We believe it is both essential and possible to make safeguards arrangements more effective. (p. 147)

Each person must decide whether he or she thinks it possible to make safeguards one hundred per cent effective. The First Report, after describing their manifest weaknesses, nevertheless clings to safeguarding agreements as the only possible means to prevent theft and weapon proliferation.

But the prospect for genuinely improved and fully effective safeguards is dim. The IAEA is poor and understaffed. It is likely to remain so. The majority of its income is spent, not on its inspection system, but on promotion of nuclear energy. It is in the international sphere what the AAEC is to Australia: a public relations exercise on behalf of nuclear energy. Its regulations cannot keep up with the spread of nuclear power, nor effectively combat the sovereign power of the nation state. International control is an illusion, for the fundamental problem is national sovereignty itself. Respect for that principle severely limits the terms of the NPT and operation of IAEA inspections. It will prevent the enforcement of any back-up agreements the Australian government may devise. Countries throughout the world are going to pursue their own national interest as they see it in particular political situations.

A utopia of international accord is not going to emerge overnight. Any moves to prevent the spread of weapons must take place within the present framework of nation states, many of whom are at daggers drawn. Within that framework the actions of a supplier state like Australia could have great impact. It would be quite reasonable for Australia to take the line that in the present state of world tension and inadequate safeguards, the only safe place for its uranium is in the ground where it would remain under effective Australian control. Such an action would be in line with new policies emerging in Canada and the USA to halt the free flow of nuclear goods and technology to other countries. It is possible that the world is now approaching the nuclear brink feared for the last thirty years; perhaps only an approach to the very edge of the precipice and a look at the abyss below can convince the human race to turn back.

It is significant that, at the NPT Review Conference in 1975 the superpowers came under much pressure and criticism from the non-aligned powers for their refusal to disarm. It is that kind of pressure that Australia could and should continue to apply. Unfortunately, during 1976 the Liberal government, reversing the previous government's policies and in defiance of its NPT obligations, abandoned the attempt to set up a nuclear-free zone in the Pacific and Indian Oceans, invited the US nuclear fleet into Australian harbours and began to develop Cockburn Sound in Western Australia as a nuclear base. This has increased tension between the USSR and the USA in the region, and illustrates what a potent effect Australian policies can have on international relations for good or ill, particularly at this critical moment in world history.

Nuclear Terrorism

Advocates of nuclear power argue that nuclear terrorism or blackmail is virtually impossible because safeguards at nuclear plants will prevent theft of materials – anyway, they say, terrorists would not have the skills necessary to fabricate a bomb using stolen materials.

Competent, qualified and objective commentators regard this view as misleading and dangerous. Mason Willrich and Theodore Taylor published in 1974 a book called *Nuclear Theft: Risks and Safeguards*.[17] Significantly one of the authors, Taylor, is a well-known nuclear physicist who was engaged in the design and building of American nuclear bombs. Willrich and Taylor point out that the *Encyclopedia Americana* now contains a most illuminating section on how to build a nuclear weapon. It is written by one 'John S. Foster, a well-known expert on nuclear weapon technology and formerly Director of the Lawrence Radiation Laboratory'. The article describes the general principles for design of nuclear explosives and gives detailed information on the chemistry and metallurgy of plutonium and uranium. Therefore the basic knowledge of how to build a bomb is freely available. Willrich and Taylor suggest that a crude fission bomb could be made by a few persons in a matter of weeks, particularly if they were willing to risk injury or death.

Clearly a terrorist or criminal group would not, on this analysis, need to kidnap (or have as a group member) a professor of theoretical physics. One young science graduate might be able to create a home-made nuclear bomb, provided he could obtain by some means the necessary nuclear material. According to a newspaper report published in the *Congressional Record* of 29 May 1976:

The AEC ran a test on itself a few years ago just to find out how easy bomb making had become. It quietly hired two young physicists with no more experience than their Ph.D. degrees, gave them access to a small computer and an unclassified library, then told them to design a nuclear weapon and predict its yield.

The two physicists had a finished weapon in six months. Their predicted yield came within 10 per cent of what their weapon would have produced had it been fired. They now work in the weapons program at Los Alamos Scientific Laboratory, where Taylor spent ten years.[18]

Such a weapon would possess a yield of twenty kilotons of explosive power, equal to that of the Nagasaki A-bomb. It would obviously have a devastating effect if exploded under a large skyscraper or in a football stadium. It could be left in a small truck in the middle of a city, or it could be made in a cellar or simply in a room of a suburban house. If there is any question, however, about the capacity of an amateur individual or group to construct a nuclear bomb, it is quite clear that for terrorist or criminal purposes it would not be necssary to do so. A dispersal effect could be obtained without nuclear explosion by constructing some device capable of spreading radioactive material over a large area of a city.

Dispersion over a wider area might be deliberately effected by the use of some kind of aerosol spray, perhaps from a helicopter or small aeroplane, or perhaps by the use of mortars, the firing of which most sixteen-year-old school Army Cadets are familiar with. If the terrorist or criminal group were to use a dispersion technique rather than construct a bomb, radioactive material from waste disposal sites rather than pure plutonium could possibly be spread. It is (quite reasonably) assumed for the purpose of such an exercise that there are many political fanatics who are perfectly willing to endanger themselves by or to die from radioactivity in furtherance of the cause they support. The spirit of *kamikaze* is still alive all over the world, often for the most idealistic motives.

The Fox Inquiry heard evidence on this question and agreed that 'terrorism is a real danger'. The Report said:

A number of witnesses treated the possibility of terrorists making a nuclear explosive device as remote, and suggested that it would be easier for them to acquire an existing military weapon. The weight of evidence available to the Commission suggests that a terrorist team could, if conditions favoured them, construct a very destructive device. The bomb-makers would have to be able to work undisturbed for weeks or possibly months, although most of the work could be done before the explosive material was acquired. They would need to call on the assistance of at least one person with the necessary laboratory skills and sufficiently well versed in the published literature dealing with nuclear explosives.

As the amount of highly-enriched uranium in use and circulation is limited, and uranium-233 is not yet being produced, the most vulnerable material for the immediate future would appear to be plutonium. As most of this is now stored in nitrate form, the laboratory would have to be able to convert it at least to plutonium oxide. Illicit bomb-makers would have to be on guard to avoid the accidental assembly of a critical mass. To avoid personal contact with toxic airborne particles, the plutonium would have to be handled most carefully in airtight enclosures. However, it was pointed out that self-preservation was not a notable characteristic of terrorist behaviour.

The evidence before the Commission suggests that a terrorist group could use reactor grade plutonium to make a bomb with good prospects of giving a yield of several hundred tonnes of TNT. Although there would be considerable uncertainty about the yield before the bomb was actually detonated, this is likely to be of little concern to a terrorist. An explosive yield of a few hundred tonnes of TNT might be sufficient to destroy a very large skyscraper, with severe loss of life. The ionising radiation released and the subsequent fall-out would also kill and injure many people.

It was also suggested that a bomb of this kind might possibly explode during manufacture, doing widespread damage, but also killing the makers. However, this risk does not detract from the real possibility of manufacture of an explosive device by terrorists to whom the risk of death might be an insignificant consideration compared with the spectacular manifestations of success, should that be achieved. (p. 153)

In a postscript to the Fox Report the Commission stated its agreement with the conclusion in the British Flowers Report that 'Plutonium appears to offer unique potential for threat and blackmail against society because of its great radioactivity and its fissile properties'. (p. 187)

Availability of Nuclear Materials

If a person wished to obtain nuclear material for the purpose of constructing a homemade bomb or dispersal device, it could be done in one of two ways: by diversion or by robbery.

Diversion means what might be called an 'inside job'. One of the main difficulties in the handling of uranium or plutonium is that such material is difficult to measure with perfect accuracy. As previously discussed it is impossible, taking into account the processes through which the material goes in the production of energy, to obtain a strict accounting of materials used. Given a range of available measurements, therefore, it could be possible for a person engaged in a nuclear enterprise to steal material a small amount at a time, over a period gathering an amount sufficient for bombmaking purposes.

This is unlikely to happen in Australia at present. However, if we do 'go nuclear', we can expect that this problem would in due course have to be confronted. It certainly may have occurred or be occurring in the USA. No matter how good a security system is, if human beings are involved, there are inevitably, over a period, going to be failures.

The Fox First Report says:

There are many materials in nuclear fuel cycles that could possibly interest terrorists. Significant materials are plutonium, uranium highly enriched in uranium-235 and uranium-233. All these can be exploited for making weapons; plutonium has additional potential as a radiological poison . . . at present, some seven tonnes of plutonium from civilian power reactors are stored in the USA. If reprocessing of oxide fuels resumes, this quantity will increase annually, even after recycling comes into operation. Other countries which plan to recycle plutonium through reactors hold stocks of plutonium. There are unspecified quantities stored in nuclear weapons establishments throughout the world and several hundred tonnes of plutonium are contained in warheads of atomic weapons under the control of armed forces. Doubts about the wisdom of recycling plutonium are widespread, because of both the dangers inherent in the routine circulation of the material and the rapid increase in recent estimates of the cost of reprocessing . . . the evidence suggests that the most serious likelihood of diversion will arise if reprocessing and recycling of plutonium are established on a large scale. This will be essential if fast breeder reactors come into regular service. (p. 152)

There are many examples of human weaknesses and lapses in security in the nuclear industry, as indicated by this report in the *Congressional Record* of 25 June 1974[19]:

In February 1973, the Atomic Energy Commission's former top security officer, William T. Riley, was sentenced to three years' probation. An investigation revealed that Riley had borrowed $239 000 from fellow AEC employees and has failed to repay over $170 000 part of which he used at the race track.

Clearly such a person could not be assumed to be immune to blandishments of bribery, for example.

However, while diversion, by juggling the upper and lower limits of MUF, is theoretically possible and, over a sufficiently long period perhaps even likely, most commentators agree that the greatest risk of criminals or terrorists obtaining the two to twenty kilograms of plutonium or uranium necessary to make a bomb lies in the act of transporting material. In April 1974, the US General Accounting Office (GAO) presented a report on AEC security. It noted, in one incident of transportation it looked at, the following deficiencies:

1. A shipment was made on a flatbed truck with an open cargo compartment; ... 2. The truck was not equipped with an alarm or communications equipment; ... 3. The truck driver was alone and unarmed; ... 4. There was no preplanned routing; the driver chose his route; ... 5. There were no periodic call-in points to let the shipper or receiver know the truck's whereabouts and to confirm that no problems had been encountered en route; ... 6. The seals on the shipping containers could be easily duplicated, thus defeating the purpose of seals which was to detect unauthorised tampering; ... 7. The material was shipped in portable containers that could be carried by one individual without the aid of mechanical handling equipment; ... 8. At an airport, the material was stacked on a dolly in an open bay area.

Since the GAO report security measures relating to the transport of nuclear materials have been tightened up. And in the two decades of nuclear operations to date there has been no report of the hijacking of nuclear materials in transit. Yet if the experience of other organizations aiming at security is considered, it is obvious that slip-ups must and do over a period occur. Looking twenty years ahead, when the number of nuclear power stations will (it seems) be multiplied ten- or a hundred-fold, and when it will still be necessary to ferry material from the ground to enrichment plant, from there overseas for use, perhaps to another country for reprocessing and return – it is highly unlikely that the kind of problem referred to in the 1974 GAO report will have disappeared. One can reasonably predict in such circumstances that hijackings of enriched uranium and plutonium will occur, and that terrorist or criminal groups will thus have at their disposal material for bombmaking.

Threats and Responses

The First Fox Report says:

A terrorist organisation might stipulate some political objective as the 'ransom' to be paid. It can be asked, for example, what the reaction of the US or British governments would be if a terrorist group threatened to explode a nuclear device in New York or London if, respectively, Israel did not withdraw entirely from the occupied territories so that a Palestinian state could be created, or British troops were not withdrawn from Northern Ireland to allow integration of the province with the Irish Republic. (p. 154)

What are the possible responses that authorities could make if confronted with a threat by a terrorist or criminal group to explode a nuclear device in, say, Sydney, if a particular demand were not met?

First, it is unlikely that the Australian federal or state governments will have taken adequate precautionary measures to prevent such an event happening at all. Perhaps a totalitarian regime might be able to foresee and forestall small-scale nuclear blackmail, but for any democratic government to be able to do so would appear likely to involve such massive infringements of personal freedom as to be politically impossible. Impossible, that is, before someone actually did threaten to explode – or exploded – a nuclear device in a large city or other vulnerable place. It would probably take such a drastic event for consideration to be given to the possible methods of dealing with such a threat or future threats.

The options would be surrender to or refusal of the demand. Surrender would involve two kinds of situation (i) where the demand could be complied with instantly – such as by the handing over of prisoners or sums of money; (ii) where the demand could not be met instantly. In the first kind of situation, there would obviously be no leeway for authorities to take action such as evacuating the population, or mobilizing civil defence authorities effectively. Unless there were available some scientific device capable of nullifying or quarantining the effect of the blast (at present mere science fiction), surrender to the demands would have to be immediate and total. The second kind of situation, however, might leave some leeway for the authorities to make a promise of future compliance which they did not really intend to keep. But presumably the clever nuclear terrorist would make such a demand secretly, under threat of instant action if the civil defence authorities or the public at large were alerted or evacuation measures taken. Such a situation might be very difficult to resolve successfully.

If the threat were of conquest, the choice of surrender would also involve the difficulty that it might produce results considered nastier than the instant death of perhaps several million people and the loss of a city or two. The authorities in charge of negotiating the demand might well consider it better to be dead than subjugated.

Assuming that no effective procedures for dealing with this kind of threat will be developed until the first time it happens somewhere in the world, the likelihood of a disastrous result (i.e., the execution of the threat) will probably be least if the demand is of such a nature that surrender to it is politically possible. Paradoxically, it would probably be preferable that a device of the kind considered here should fall into the hands of a callous and brutal criminal gang, such as one organized on the profits from drug-running, than into the hands of a high-minded, idealistic, political organization such as the IRA, the Croatian Liberation Movement, or the Black Panthers. Straightforward financial demands would no doubt be easier to meet than would emotionally loaded political ones.

Refusal of the kinds of demands we are considering here must, therefore, involve a probability that the threat might actually be carried out. We may soon be looking back with nostalgia to the Cold War period of the 1950s and 1960s, when nuclear threats between major nations were mainly bluff and brinkmanship, as an era of stability. Even John Foster Dulles was aware that swift and terrible retribution against the USA would have followed the unleashing of a nuclear attack on the USSR. Besides, disaffected groups with little to lose simply cannot be threatened with retaliation, which makes it likely that an excessive demand met by a refusal will result in the execution of the threat.

One of the major problems confronting a government faced with a threat of nuclear blackmail is in deciding whether or not the terrorist threat is a hoax or genuine. In December 1976 publicity was given to some letters that had been sent to the Australian Prime Minister and some other people demanding, under threat of a nuclear explosion, that Australia should not export uranium. Newspaper reports indicated that the letters were signed 'Group of Six'. A deadline was set for an announcement by the government that it would not export uranium, but this passed by without any explosion. Apparently the threat was a hoax. The letters were not released, so possibly they were internally not plausible, but undoubtedly the government and its advisers took a calculated risk. Presumably the threats were not accompanied by other evidence about the possession of fissile material, and it was assumed there was none in the possession of whoever sent the letter. The First Fox Report gave consideration (in advance of this particular incident) to the question of 'threat plausibility':

> In any situation in which a threat was made to set off a nuclear device, the security forces and the government concerned would face an obvious dilemma; how seriously should they take the threat? On the one hand, there would be the incalculable consequences if the threat materialised. On the other, the authorities would be aware of the need to hold out against the possibility of bogus threats. It would be very easy for any individual or disaffected group to advance demands on the pretension of a nuclear threat.
>
> Some witnesses suggested that, to an extent, the authorities could rely on the supposition that, as the terrorists presumably would wish to achieve their objective, they would be prepared to offer evidence of the existence of their explosive device. Examples given of the form this evidence might take include a sample of the plutonium used, or details of the plutonium and how it was obtained, and of the bomb and how it was made.
>
> Production by the terrorists of some plutonium would be much more convincing evidence than the production of technical details, although it is conceivable that a group which had produced a bomb might not be able to provide excess plutonium on demand for logistic reasons or, alternatively, that a group which had not produced a bomb could provide some plutonium if any had in fact been diverted. Design features produced would almost certainly be copies from available literature whether or not the bomb actually existed. It is concluded that the authorities probably would be in no better position to judge after the information or material was given than before whether the threat was real. Similarly, details supplied about the plutonium and the method used to obtain it would probably not enable the authorities to determine in the time required whether a clandestine diversion had in fact occurred. If the alleged source of the plutonium was in another country, there would be even less chance of making a reliable assessment of the likelihood that the threat was genuine. (p. 154)

As time goes by, if the quantity of nuclear materials in the world increases, the difficulty of deciding whether a threat is genuine or is a hoax will increase; ultimately the decision will be not so much a calculated guess as out-and-out Russian Roulette. Again, to quote the First Fox Report:

> There is a very real risk that the opportunity and the motive for nuclear blackmail will develop with time. Some common characteristics of terrorist groups suggest that they might attempt to make and explode atomic bombs or make other terroristic uses of nuclear materials or facilities. These include lack of concern for their own safety and for the suffering caused by their attacks. Since they have no territory of their own, fear of nuclear retaliation is not an inhibiting factor. Major difficulties could arise in attempting to determine the reality of a threat by a group to explode an atomic bomb, to spread radiation from a reactor, or to disperse plutonium. Either acceding to or refusing the demands of such a group could have very adverse consequences for society.
>
> Measures designed to prevent theft of nuclear materials and attacks on nuclear installations have been tightened in recent years. Welcome as those measures are, the evidence is that the risks are presently real and will tend to increase with the further spread of nuclear technology. (p. 159)

Policing the Plutonium Economy

In order to stop uranium and plutonium from getting into the wrong hands, extreme security measures would have to be taken. A nuclear society, with adequate safeguards against theft of nuclear materials or accidental exposure of population to waste products, would require a large measure of social control. First, there must be strict surveillance of personnel involved in the industry. According to American lawyers, security measures could involve the legal use of informers and wiretapping. In a recent article, 'Policing Plutonium: the Civil Liberties Fallout', Russell W. Ayres states:

> The urgent need to prevent thefts of plutonium will lead to a loosening of standards for government conduct of covert surveillance. The government will probably take full advantage of the broad powers which the courts have allowed it in the use of informers and infiltrators. Moreover the case for using wiretapping to uncover plots to steal plutonium is very strong.[20]

A strengthening of safeguards, of physical security and of personnel surveillance would result in a highly controlled society in which rights of free speech and privacy would become memories of a former age. The numbers of armed security officers would multiply, as well as their powers to deal with suspects. Russell Ayres, in the same article, discloses that it has already been proposed in the USA that a special nuclear police force should be formed with particular powers. It is therefore difficult to see how a nuclear society could maintain the free institutions and individual rights common to western democratic countries.

These measures would be necessary in what are commonly called 'stable' countries, but an even greater problem is posed where there is the possibility of civil unrest. To gain control of the state's power source is an obvious strategic aim of a revolutionary plot. Whether a new regime would or could successfully maintain safeguards on those utilities is another matter. In the case of a spontaneous violent revolution there is the additional possibility that technicians would be killed, temporarily stored wastes accidentally released, and a complex power plant fall into the hands of insurgents. Areas where large-scale nuclear development is planned that could fall into this category include southern Africa, Latin America and the Middle East. Since revolution could result in the loss of expert control over nuclear facilities and wastes, authoritarian regimes would be likely to increase and justify their already stringent security and repressive measures. In this way, nuclear power would encourage the trend to authoritarianism and stimulate, in reaction, violent revolution.

The most frightening aspect of the nuclear society, however, is the secrecy that would surround the daily operation of the state's power supply. Those in control of security measures could not be subject to public scrutiny. To give one recent example: Australians were assured by their government that adequate precautions would be taken against an accident to nuclear-powered US warships in Australian ports. However, the Prime Minister refused to table the full Defence Department report in Parliament because it would be 'contrary to the interests of Australia to publish it'. Parliamentarians and the public, therefore, can have no means of judging whether safeguards are adequate or what kind of covert surveillance they may involve.

The general result in a nuclear society of lack of public scrutiny would be that a small group of technocrats would become the real governing force. Because the fuel for domestic energy would also be a strategic material, to go nuclear could mean a qualitative change in governmental institutions. Real power could reside with the controllers and operators of nuclear power. Parliament, unable to enter into open discussion of the operation and protection of the nation's power source, could become a cipher.

This aspect of the effect of nuclear power has just begun to receive attention. But it is a major consideration and one that outweighs any possible short-term material benefits to society that nuclear power may be claimed to convey. In most countries there are laws relating to atomic energy already on the statute book or being rapidly legislated which give even democratic governments unprecedented powers. In Australia, for example, the Atomic Energy Act, passed in 1953 at the height of the McCarthy period and not amended in substance since, provides both a deterrent to public inquiry into the activities of the AAEC and an attack on civil rights.

The crucial clause of the Act is Section 47 concerning proof of intent to prejudice the defence of the Commonwealth of Australia. This section refers to prosecution of persons under Sections 44, 45 and 46. These sections make it an offence, punishable by imprisonment for twenty years, for a person to disclose restricted information with intent to prejudice the defence of the Commonwealth. 'Restricted information' is widely defined in Section 5 as information concerning 'prescribed substances' which are in turn defined, widely, as 'uranium, thorium, plutonium, neptunium or any of their respective compounds'. In other words, any kind of information about uranium and its products which has not been officially released by the AAEC becomes restricted. This is a wide definition indeed and makes the severity of the security clauses even more acute.

This is particularly so in view of Section 47. Section 47a rules that it is not necessary to show a person guilty of a particular act prejudicing the defence of the Commonwealth if 'from the circumstances of the case, his conduct, or his *known character* as proved, it *appears* that he acted with intent to prejudice the defences of the Commonwealth' [our italics]. This clause is in absolute contradiction to usual procedures. A person could easily be victimized under this clause, because of his political opinions, for example, since the court does not have to prove that he was guilty of a particular act. All the court has to find is that the person appeared to act in a subversive manner. The penalty is imprisonment for twenty years.

Section 47b states that, if a person has been proved to have disclosed restricted information, 'it shall be deemed, unless the contrary is proved, to have been so dealt with with intent to prejudice the defence of the Commonwealth.' Therefore the accused is presumed guilty until he proves his innocence and the burden of proof is thrust on him.

Other security clauses worthy of mention are Sections 50, 54, 58, 62 and 63.

Section 50 permits arrest and search without warrant on mere suspicion. Yet Section 54 states that no action can be taken against the Commonwealth or the Commission in respect of wrongful arrest, detention or search. Compensation for arrest without reasonable cause is at the pleasure of the Governor-General.

Section 58 widens the definition of an offence by saying that 'a person who does an act preparatory to the commission of an offence against this Part [Part IV Security] is guilty of that offence.' What comprises 'an act preparatory to the commission of an offence' is *not* defined.

Section 62 allows for secret trial. If the judge or magistrate considers it to be in the interests of the defence of the Commonwealth he or she can exclude the public from the court, order that no part of the proceedings be published and order that affidavits, information or any document may be expunged from the records of the court.

It should be noted, however, that by Section 63 the consent of the Attorney-General must be given for an action to proceed, although a person can be arrested and detained pending the consent of the Attorney-General.

It is clear that the security clauses of the Act threaten the rights of the individual. They provide for summary arrest, secret trial, destruction of the record of evidence, with the burden of proof of innocence on the accused who cannot take legal action for wrongful arrest against the Commission, who can be arrested and convicted simply on grounds of his 'known character' and, if convicted, can be subjected to a long term of imprisonment. Such is the protective barrier that shields the AAEC.

In talking about the police-state methods necessary to safeguard plutonium, we are not speaking of a distant future. It is happening now, wherever nuclear power is established. In 1976, for example, the British parliament passed the Atomic Energy (Special Constables) Act. The Act and its implications were described in a short but highly significant book published later in that year by Michael Flood and Robin Grove-White for Friends of the Earth, the Council for the Protection of Rural England and the National Council for Civil Liberties. It is called *Nuclear Prospects* and its content is spine-chilling. It describes how parliamentary control over Britain's nuclear industry has already been eroded, how members of parliament cannot, for security reasons, inquire about the movement of plutonium on the public transport system of their densely populated country. It predicts that the security measures now being taken are but a foretaste of what would be required should Britain develop fast breeder reactors.

The Special Constables Act forms an historic break with England's democratic and police traditions. The Act confers on a special police force under the control of the Atomic Energy Authority the power to carry arms, which the regular police do not, and to arrest on suspicion. It is in effect a private army. Yet, as Flood and Grove-White say in their book: 'Few would doubt the necessity for such powers for guarding plutonium and similar materials'. (p. 20)

The bill became law after many regrets were expressed in Parliament that it went 'against the grain for anyone with our tradition', as Energy Minister, Anthony Wedgwood Benn said. Yet in the face of plutonium theft, civil liberties had to be curtailed.

The full significance of the Special Constables Act and further security measures that would be necessary in a plutonium economy is that they remove governments and their atomic energy departments from public scrutiny, give those institutions the power of general search, phone tapping, infiltration of citizens organizations, arrest on suspicion and/or on grounds of 'known character' – all the apparatus of totalitarian government. These powers could be used for any purpose, to suppress a political minority, to silence criticism of the government. Such powers are an attack on rights of individual privacy, freedom of speech and freedom of association which have been painfully built up in democratic countries after centuries of constitutional development.

The choice, should nuclear power go ahead, is between a world in which many nations and terrorist organizations are armed with nuclear weapons, and a world in which every nation, to keep its plutonium secure, is a police state. Probably both would occur. Democratic institutions would be eroded by security legislation without preventing, in the end, the theft of plutonium. This is the nuclear prospect. This is what Australia would be helping to create should it export uranium.

6 A Nuclear-Free World
Jeffrey Nicholls
Michael Bell

Western civilization needs large amounts of energy to survive in its present form. This energy comes from the fossil fuels – coal, oil and gas – all of which will be exhausted in the foreseeable future. We do not have a lot of time to develop new sources of energy to replace the fossil fuels. As this fact has become more obvious, more and more public discussion has centred on energy, particularly nuclear energy, which is supposed to be our saviour. The dangers of nuclear power to the world's natural environment and to human freedom and social development have been described elsewhere in this book. There are other problems too. The world's resources of material from which to generate nuclear fuel are also finite, so that nuclear power would be at best a temporary solution to problems of energy supply. The world's resources of capital can hardly afford to provide the huge investment needed to replace fossil energy with nuclear power. Finally, nuclear power is totally unsuited to most human energy needs in both the developed countries and the less developed countries.

Nuclear power is not the answer: we should look for others. The science and technology of solar energy have been almost ignored in the past by a world fascinated by the power in the atom. A mythology has grown up about the difficulty and expense of using solar energy – much of which turns out, on close examination, to be unfounded. Solar energy not only provides the key to solving our energy problems permanently, it also allows us to lay the groundwork for a comprehensive and lasting solution to the environmental problems that are destroying our small planet and shortening all our lives.

Human beings are now realizing that they all live in one small world; environmental changes made by one person or industry in one part of the world eventually affect everyone. Chemicals released at one point on the globe eventually find their way to other places, even the most remote. Pollution of the environment must be controlled for the human race to survive. There are two basic approaches to pollution control. One is simply to add pollution control devices to existing plant. We can put exhaust emission controls on our cars, run all the stormwater from our filthy streets through huge water purification plants, and try to develop infallible methods of storing our nuclear wastes. In the long run such control measures will become prohibitively expensive and complicated. The second approach is to convert to a technology that is based on the same energy and material flows as the natural world. The sun is the natural source of energy. Once we have turned to the sun for our energy, we have made the first step towards a technology in harmony with the natural world.

Energy Policy in a Democracy

Politics is the process of deciding what to do. It takes various forms – from discussion and argument between members of families and friends, to parliamentary debates (which all too often are ritualized slanging matches) through to elections and referenda.

Over the centuries it has been accepted in theory that political decisions should be made democratically. Democracy means literally *people power*. Democratic politics means that everybody affected by a decision has a right to participate in making it: nobody should be deprived of the right to influence a decision that affects him. But theory is one thing and practical politics quite another. We have a very long way to go before our systems become fully democratic. At the moment many, perhaps most, of the decisions affecting the lives of ordinary people are made by anonymous bureaucrats, experts, commissions, secret inquiries, company boards, marketing cartels and other undemocratic institutions. People are excluded from involvement in decisions affecting them, by secrecy, ignorance, feelings of futility and even laziness.

Energy is a matter of worldwide importance and, like all matters concerning resources and the environment, concerns everyone on the planet. Ideally, everyone should be allowed to contribute to energy policy decisions. On a worldwide basis, we are far from the ideal situation for this, but a small nation with good communications like Australia should be able to involve a large proportion of its population in the decision-making process.

Democracy is not just a matter of voting. A democratic decision has three stages: the discovery and general publication of the facts of the matter, public discussion and debate on these facts and their implications, and some sort of counting of heads to see what the consensus of opinion is. Anyone concerned to thwart the democratic process can attack it at any of these three stages by obscuring or distorting the facts or preventing them from being published, by avoiding public debate where all shades of opinion are given a hearing, and by giving no opportunity for a fair vote on the issue.

Clearly, there are many in this country who would like to prevent a proper public decision on uranium and nuclear power. Mr Justice Fox and the other Commissioners who conducted the Ranger Inquiry discovered this, saying, in the First Report that 'a few of the government officials who appeared before us showed a reluctance in communicating matters of importance to the Commission' (p. 6).

Since this Report was released Australia has seen a completely farcical parliamentary debate on the uranium issue and a lot of statements by the mining lobby and others that we have had the Inquiry and should now get on with the mining. Yet the stated purpose of the Fox Inquiry was not to make a decision on uranium mining but to determine the facts as a foundation for informed public debate. In the words of the Commissioners:

This First Report of necessity deals with some very broad issues, with respect to which different minds can quite readily come to different conclusions. Having regard to this, and the limitation on the ambit of the inquiry, the view might be taken that, in relation to those issues, *our findings of fact will prove more valuable than our final recommendations.* [our italics] (p. 5)

The final and most important recommendation of the Commission in their First Report relates to the second stage of the democratic process:

Our final recommendation takes account of what we believe to be the policy of the Act under which the Inquiry was instituted. It is simply that there should be ample time for public consideration of this Report, and for debate upon it. We therefore recommend that no decision be taken in relation to the foregoing matters until a reasonable time has elapsed and there has been an opportunity for the usual democratic processes to function, including, in this respect, parliamentary debate. (p. 186)

No decision, even when taken with the best available information and full public participation, is final. The world does not stop still; as well as social and political changes, there are scientific and technological discoveries which open up new possibilities. Decisions must always be open to review as new information comes to hand and our understanding of our relationship to the world changes. For this reason it is important not to make any irreversible commitments which may turn out to be wrong in the long run. Taking such changes into account, the First Report recommends that 'Policy with regard to the export of uranium should be the subject of regular review'. (p. 186)

The Fox Reports and the British Flowers Report are two of the first major independent investigations of nuclear power in the thirty years it has been in existence. Most other reports have been prepared within various atomic energy agencies and contracting companies and are, for this reason, not very reliable when it comes to objective evaluation of the costs and benefits of nuclear energy. One can hardly imagine a body such as the Australian Atomic Energy Commission, for instance, seriously considering the possibility of recommending its own dissolution because atomic energy is too dangerous.

Anyone seriously interested in nuclear energy must read these two reports, which contain a tremendous amount of information on nuclear energy and do not minimize its dangers as the nuclear industry has done. The Commissioners who prepared the Flowers Report admitted, 'We are sufficiently persuaded by the dangers of a plutonium economy that we regard this as a central issue in the debate over the future of nuclear power'. (p. 186) And yet they found that the nuclear industry has almost ignored the danger. The Flowers Commissioners found it remarkable that none of the official documents they saw in their study of nuclear power conveyed any unease about the creation of plutonium on a large scale. Their finding was that,

The dangers of the creation of plutonium in large quantities in conditions of increasing world unrest are genuine and serious. We should not rely for energy supply on a process that produces such a hazardous substance as plutonium *unless there is no real alternative* [our italics]. (p. 193)

All nuclear reactors produce plutonium, and there are real alternatives. Unfortunately very little is known about these alternatives. Technological information costs money and it costs money to distribute information. Tens or even hundreds of billions of dollars have been spent on research and development of nuclear power and the population at large has been subjected to endless propaganda about its supposed virtues. Solar energy has been almost completely ignored. Only very recently has expenditure on solar energy research been raised to a significant fraction of the expenditure on nuclear energy. In Australia annual research expenditure on nuclear energy is at least twenty times as great as that on solar energy. In other countries the position is worse. Australia's contribution to the world development of nuclear energy has been almost negligible, despite the expenditure. On the other hand, Australia was, until the last few years, one of the leading nations in solar energy research. We are now falling sadly behind, as witness after witness has pointed out to the recent Senate Standing Committee on National Resources Inquiry into Solar Energy.

This means that the energy debate cannot really get under way until the alternatives are subject to serious study. A political decision to facilitate this study must be made. If it is not, the wrong decision on the world's future energy supplies may be made out of ignorance. The Fox Inquiry heard a great deal of evidence concerning solar energy, most of it from people without the financial, institutional or information resources to make a strong case for solar energy. The First Report found that:

Unfortunately most of the evidence (on solar energy) both from those taking an optimistic view of the potential of solar energy and from those taking a pessimistic view, was expressed in a form which makes it difficult to assess the likely contribution of the various solar energy technologies to future requirements. (p. 41)

The Fox Inquiry cost $800 000; the case for nuclear energy was put with the full resources of the Australian Atomic Energy Commission (which is financed by taxpayers' money), a number of large mining companies, and the overseas information resources upon which these bodies can draw. It is a pity, then, that no public money was allocated to make the case for the alternatives to atomic energy. In a real sense the Australian public has not had true value for the money spent on the Inquiry because the case against nuclear energy was not able to be fully put.

If the decision about energy is to be democratic money must be spent to get the information. Funds must be made available to employ scientists, technologists, economists and all the other necessary experts to supply the information necessary to make a wise decision. Billions of dollars have been spent on nuclear power and it is still surrounded by probably insoluble problems. Before it can honestly be said that there are no real alternatives, similar resources should be devoted to solar energy. Until the world has spent perhaps one thousand billion dollars on solar energy and found it wanting, no one will be in a position to decide finally that there is no real alternative to the plutonium economy.

Planning an Energy Economy

Little is known, comparatively speaking, about the alternatives to nuclear energy, but there is not complete ignorance of the possibilities. There are only four known sources of energy capable of satisfying the world's ever growing appetite for energy: fossil fuels, upon which we currently rely, nuclear fission, nuclear fusion and the sun. Even at present rates of consumption, world reserves of fossil fuels will only last a few centuries. If consumption continues to grow, their lifetime will be even shorter. Few can dispute this and everyone seems to agree that alternatives should be sought now. It takes a long time to invent, develop and widely disseminate a new technology – probably at least fifty years. The world has been aware of the possibilities of nuclear fission for nearly forty years. Last year nuclear fission provided about one half of one per cent of the world's energy needs. Even the most optimistic estimates for the year 2000 show nuclear energy providing only about 10 per cent of the world's energy, nearly fifty years after the first nuclear power station began to operate.

Nuclear fusion (the production of energy by fusion of hydrogen atoms) is quite a remote possibility. It is still a matter of theory and laboratory experiment, even though hundreds of millions of dollars have already been spent on it. Experimenters seem to expect a successful laboratory demonstration of nuclear fusion in the near future. When this does occur, fusion energy will be in the same position as nuclear energy was in about 1938, with a long way to go before it has any hope of supplying energy to the world's economy.

Solar energy is as old as the earth, or older. The sun is an average sort of star, 150 million kilometres from earth. It is 1.4 million kilometres in diameter, has a surface temperature of about 5500° Celsius, and radiates an immense amount of energy, mostly as light visible to the human eye. Solar energy arises from the fusion of hydrogen nuclei to form helium in the intensely hot interior of the sun.

The earth intercepts less than one billionth part of the energy leaving the sun, yet this is equivalent, in a day, to 600 000 million tonnes of coal, roughly the world's total recoverable reserves of this fuel. Of all the energy reaching the earth's ecosystem we are only using 1/25 000th of its potential. Astronomers agree that the sun has been providing the earth with energy since it was formed four or five billion years ago and it seems likely that the sun will continue to shine with its present strength for another five billion years.

Only nuclear fusion and solar energy have any hope of providing a long-term solution to the world's energy problems. Nuclear fission, based on uranium, is only a temporary energy source at best. According to the First Fox Report, the world's currently known reserves of uranium are 3 850 000 tonnes. Estimates of the uranium needed for reactors planned to the year 2000 range between 2 200 000 and 4 300 000 tonnes. From these figures it is apparent that it will not be long before there is a shortage of uranium. Nuclear fusion, if it works, will be based on the elements lithium and deuterium. Sufficient quantities of deuterium are available in the ocean to supply the world at current rates of consumption for billions of years.

Solar energy already provides a significant proportion of the world's energy needs, although this does not show up very clearly in official statistics. Hydro-electricity is a form of solar energy since it is the sun's energy which lifts water up to the high dams that power hydro-electric turbines. In Australia, hydro-electricity provides about 17 per cent of total electricity generated and 2 per cent of the total energy used in the country, which includes coal and oil as well as electricity. Apart from direct radiation, solar energy appears indirectly in the form of wind, waves, temperature differences in the oceans, and high lakes and rivers that can be used to generate hydro-electricity. The other major renewable sources of energy are tides and the heat within the earth. The earth's heat (geothermal energy) can be used directly for heating, or converted into electricity. The other forms of energy need to be converted into electricity before they can be used.

The sun also provides us with food. The Australian wheat crop contains about the same amount of energy each year as all the electricity generated in the country. Our sugar production contains about 20 per cent of the energy in the wheat crop. The total energy contained in food produced in Australia is roughly 10 per cent of the energy consumed in coal and oil. Australia is a highly developed country with a high usage of fossil fuels. In less developed countries the energy consumed in food often far exceeds the energy consumed in fossil fuels.

Solar energy also helps us in many ways that are not recorded in any statistics at all. It dries clothes, hay and fruit. It warms the earth, purifies and distils water, and causes the winds which distribute energy around the earth, purify the air over cities and drive windmills. When people speak of solar energy, they usually mean energy from the sun used in such a way as to replace some present consumption of electricity or fossil fuel. For instance, a solar water heater replaces an electric water heater, thereby saving some electricity or fossil fuel. There is a sharp distinction in our thinking between energy that is counted in statistics and energy that is not counted: the electricity used to power a clothes-dryer is counted, whereas the solar energy used to dry clothes on a line is not counted. This is a result of the way statistics are collected. Data are collected, often, only on those things given monetary value in the economy. Solar energy is free, so, like the unpaid work done by women in the home, it's taken for granted.

The amount of energy used for man's technological purposes – that is, energy used in machines – is tiny compared to the total amount of solar energy flowing from the sun through the earth's natural systems. All this unaccounted solar energy is not wasted: directly or indirectly, it provides us with all the basic necessities of life – light, air, warmth, water and food. The natural flows of energy in the world are huge compared to the man-made flows. The key to human survival is to integrate our own technological energy usage with the natural usage of energy all around us.

Why has this not been done? It is hard to tell causes from effects. The natural world is a vast integrated system. Human technology has grown up in bits and pieces within this system. Most technological developments have been for a particular purpose at a particular time – ships are to carry goods over water, tunnels are to get through mountains and electricity has been found to be the most convenient way to supply energy to homes and factories. Generally the easiest way to solve a problem is to build on what has gone before; it is difficult to go back to scratch and start again. Steam-powered cars might have many advantages but so much effort has gone into the internal combustion engine that it is unlikely to be displaced by steam for a long time to come. The development of technology has been strongly influenced by expedience, economics and the historical accidents of discovery. Often the results have not been good. The cheapest and most convenient chemical factory might also be the most polluting and dangerous.

The growth of scientific knowledge and technological know-how should have put us in a position to go about the development of technology in a slightly more orderly manner. Technology is a means, not an end. Its purpose is not killing people or winning wars or making money. The only legitimate purpose for any form of technological device is to increase the quality of human life by providing food, health, communication, recreation or something else of value to humanity. Every form of technology must be examined in this light. If it detracts from the overall quality of human life more than it adds to it, then it is destructive to life, even if it does make money for a particular entrepreneur, or helps a particular government to win a war.

Once, the costs and benefits of a particular industry were found by trial and error. Lead-based paints, for instance, were found to be better and cheaper than earlier paints. Only later was it discovered that lead-based paints contributed significantly to lead poisoning, particularly in children. Once the connection between lead-based paint and lead poisoning had been established, there was still the political and economic problem of preventing the manufacture of more lead-based paint, and replacing all such paint on roofs, furniture and toys with safer products. This still has not been completely done.

Scientists can now often predict fairly accurately in advance whether a particular product will be helpful or harmful. The scientific and technological means exist to do this, but the political mechanisms to make sure they are used have still to be developed. It is still possible for the board of a plastics company to decide to market a product that has not been proven completely safe, although, as people become more aware of the dangers, it will become harder and harder to do so.

Energy is the most vital commodity of all the products of technology except food, and energy technology is probably the least subject to political control. The coal and oil resources of the earth are under the control of quite a small number of huge international companies which have sufficient financial and political strength to ignore or strongly influence all the national governments in the world. Any attempt to work out the best energy policies for the world must take the power of these companies into account. They have decided on nuclear power as the successor to fossil energy, a view that is very much in line with the conventional or official theories on the development of energy resources. Since the fifties it has been generally accepted in government and energy industry circles that nuclear power was the logical successor to fossil energy. This idea has gathered such momentum that even now, when the nuclear power industry is in serious difficulty, faith in its future remains unshaken and industry spokesmen continue to predict a prosperous and rosy future.

The only ground for their faith is the belief that the political strength of the nuclear power companies and governments which have committed themselves to nuclear power will be sufficient to bulldoze through the opposition despite the manifest technological, environmental and economic failings of nuclear energy. Even this ground for their faith is weakening, as the facts about nuclear power become more widely known and more and more people, horrified by its dangers, come to oppose it. The dangers of nuclear power have been discussed in the preceding chapters of this book. The purpose of this chapter is to demonstrate the advantages of solar energy as an alternative to nuclear energy, for, the dangers aside, nuclear power has no hope of replacing fossil fuels as our technological energy source.

The first reason for this is the limit of the nuclear energy resource. It has already been pointed out in this chapter that there is likely to be a shortage of uranium at the end of this century, when nuclear power is supplying only about 10 per cent of the world's energy needs. Proponents of nuclear power pin their faith on the breeder reactor which will not only be able to get about fifty times as much energy from a given amount of uranium, but will also make it economic to mine the huge resources of extremely low-grade uranium ore that are known to exist.

The development of breeder reactors has run into many difficulties and it seems unlikely now that they will make any impact on nuclear energy production or the consumption of uranium before the year 2000. Even then the growth of the breeder reactor population will be slow. Ordinary reactors have to be operated for a long time to produce the plutonium needed to fuel each new breeder reactor, and a breeder reactor must also be operated for a long time to produce new fuel for another reactor. Above all, breeder reactors are going to be very much more expensive than conventional reactors. Cost is a major cause of the problems currently facing the nuclear industry.

The cost of nuclear reactors is increasing at an alarming rate. In the USA a typical large (1000-megawatt) reactor begun in 1967 was completed in 1972, having taken five years to construct at a cost of about US$140 million (A$130 million). A similar reactor ordered in the USA in 1975 will not be completed until 1984, nine years later, at a cost of US$850 million. The cost of the actual reactor is only part of the story. Nuclear reactors can only be used to generate electricity, which must be transmitted to the customers through power lines, transformers, meters and other equipment. Transmission and distribution of electricity costs more money than its generation. In New South Wales, for instance, the cost of distribution is twice the cost of generation, and the capital cost of the equipment needed to get electricity to the customer's meter from a 1000-megawatt power station was about $160 million. To this must be added the cost of meters and wiring in the customer's own premises, which would be another $500 million at the very least.

Taking into account the current price of reactors, and the fact that because reactors are unreliable and electricity must be generated at the same time as it is used, the whole system can only be used at about 50 per cent of its capacity. On the average, it turns out that about $2200 million has to be expended to supply 1000 megawatts of electricity from a nuclear power station to the equipment that is to use the electricity. This figure does not include the cost of the equipment itself. In comparison, the Gas and Fuel Corporation of

Victoria employs about $140 million worth of capital equipment to deliver 1000 megawatts of gas energy to the appliances that use it, and the oil industry needs only about $80 million worth of capital equipment to produce, refine and deliver 1000 megawatts of oil energy to the point of end-use. This is more expensive than oil elsewhere in the world because Australia's oil industry is small and the oilfields are offshore, where production costs are very high compared to oil wells on land.

The all-nuclear energy economy will be an all-electric economy. It is doubtful whether any country in the world could afford to convert to such an energy economy. If it were planned to convert Australia completely to nuclear power over a twenty-five-year period, approximately 50 per cent of the nation's gross fixed capital expenditure would have to be devoted to the task, on the simple assumption that energy consumption, gross fixed capital expenditure and the cost of nuclear plant remained constant over the period. It is also assumed that none of the plant would have to be replaced over that period. If it did, the proportion of capital expenditure required would increase. Twenty-five years is the approximate lifetime of a nuclear reactor. At present about 7 per cent of the nation's capital expenditure is devoted to energy production and distribution equipment. Obviously to convert to nuclear energy would be an incredibly expensive process and would starve the rest of the economy of much needed capital. The conversion to nuclear power would be even more expensive if breeder reactors were used. If they are not used, the immense expenditure on nuclear power will provide only a temporary source of energy because of the shortage of uranium. If they are used, nuclear energy is likely to absorb a huge proportion of the world's capital for a long time to come.

The shortage of capital is already making it very difficult for the nuclear industry to achieve the growth rates it had hoped for. Capital is attracted by high rates of return and security of investment. The nuclear power industry can offer neither of these. The security of capital invested in the nuclear industry is threatened because there is an ever-increasing chance that reactors will become socially and politically unacceptable before they have lived out their design lifetime, even if they are not closed down by major or minor malfunctions.

Nuclear power is capital intensive. Approximately three quarters of the cost of nuclear electricity is the cost of capital. In New South Wales the average cost of generating electricity from coal-fired power stations is about 0.8 cents per kilowatt hour. The capital charges alone on a nuclear power station come to about 1.2 cents per kilowatt hour if it is to pay itself off over twenty-five years and return 10 per cent on the capital invested. If it is to be competitive with the coal-fired power station, the return on investment can be only 5 or 6 per cent, not a high enough return by today's standards. It was once thought that nuclear electricity would be much cheaper than electricity generated with coal, but this no longer appears to be so. The rapid increase in the capital cost of nuclear plant is gradually giving fossil-fuelled electricity generation a bigger and bigger advantage over electricity generated by nuclear power.

Solar energy is also capital intensive. Approximately 90 per cent of the cost of solar energy comes from the capital cost of equipment. The usual argument put up against solar energy is that it is too expensive. True, it is more expensive than oil or gas energy, but it is generally cheaper than electricity produced in coal-fired power stations and is likely to be much cheaper than electricity generated from future nuclear power stations.

Solar energy enjoys two massive advantages over nuclear energy. The first is that it is already distributed fairly evenly over those parts of the world where most of the population lives. Nuclear power stations can only produce electricity that is expensive to distribute: the cost of moving the electricity from the power stations to the device where it is finally used can be up to twice the cost of generating the electricity. In most applications, solar energy can be collected at the point of use, avoiding distribution costs.

The second advantage is that solar energy can be collected in the form that is needed: it is not necessary to convert it into electricity. To heat water with nuclear energy, the heat from the reactor must first be changed into electricity, then the electricity transmitted to the electric water heater in the customer's house where it is changed back into heat. Solar energy can be collected by absorber panels directly as heat, completely bypassing the difficulty and expense of converting it to electricity. Critics, when comparing the cost of solar energy to other energy sources, overlook this fact, believing they can prove solar energy uneconomic simply by showing that it costs more to generate electricity from the sun than from fossil or nuclear fuel. In doing this they overlook one of the greatest advantages of solar energy.

Electricity is at present a relatively unimportant and very wasteful component of the world's overall energy supply. In 1974 electricity provided about 12 per cent of the energy used by final consumers in Australia. The electricity generation and distribution process is inherently very wasteful. In New South Wales in 1974 only one quarter of the energy contained in coal burnt in power stations finally reached the customers' houses as electricity. Most of the energy loss was released into the environment as waste heat at the power station. Nuclear power stations waste even more heat than coal power stations, so that the all-nuclear economic energy economy would have immense problems disposing of waste heat without causing very great damage to the environment.

Amory Lovins, in his article 'Energy Strategy: The Road Not Taken' (*Foreign Affairs*, October 1976) gives some figures for the USA that do not seem to differ very greatly from what might be expected for Australia. Fifty-eight per cent of all energy in the USA is required at the point of end-use in the form of heat, roughly half above 100°C and half below. Another 38 per cent of energy provides mechanical motion – 31 per cent in vehicles, 3 per cent in pipelines and 4 per cent in industrial electric motors. The rest, a mere 4 per cent of all energy delivered to consumers, represents *all* lighting, electronics, telecommunications, electrometallurgy, electrochemistry, arc welding, electric motors in home appliances and in railways, and similar end-uses that *require* electricity. Any further use of electricity is wasteful of both energy and capital.

Despite this, energy forecasters predict that the proportion of energy used as electricity will increase. In the UK the 'official strategy' as determined by the Flowers Commission sees the use of electricity increasing from 13 per cent in 1975 to 30 per cent in 2000 and nearly 60 per cent in 2025. At this point energy losses in the fuel industries, mainly the electricity-generating industry, would be one and a half times the total energy actually consumed by end-users. This absurd wastefulness would have been forced on society because nuclear power can only be used to generate electricity.

Solar energy can be used as low-temperature heat considerably more cheaply than electricity. A typical low-temperature solar heat collector used to produce domestic hot water costs about $100 per square metre. In typical Australian conditions, about eight square metres of collector are required to produce a kilowatt of low-temperature heat energy, if averaged over day and night, summer and winter, and a typical figure for collector efficiency is adopted. In other words, about $800 needs to be invested to obtain an average of one kilowatt of power. This is roughly the same as the investment in New South Wales and Victoria (excluding meters and house wiring). Assuming that such a collector has to pay itself off at 10 per cent interest over ten years, and adding a bit for maintenance costs, the cost of energy comes out at slightly less than 2 cents per kilowatt hour. This is a little more than the price of off-peak electricity, but about 20 per cent less than the average cost of generating and distributing electricity in New South Wales and Victoria.

Thus, if it were not for off-peak electricity, solar water heating would be cheaper than electricity. The price of electricity is worth examining. In general, electricity prices seem to be artificially low because in Australia the electricity industry is publicly owned with access to very cheap capital and fuel. In New South Wales, for instance, the cost of electricity generated and distributed to the customer's meter averages about 2.5 cents per kilowatt hour. The cost of the coal used averages about $5 per tonne and the average interest payments on the capital employed is 3.4 per cent. If the electricity industry had to pay world prices for its steaming coal (about $20 per tonne) and return 10 per cent on its capital investment, the cost of electricity would jump to about twice its present level, and the advantage of solar energy would become very clear. At present, solar water heaters, financed by expensive private capital, have to compete against artificially low electricity prices and even lower off-peak tariffs.

The electricity industry claims that off-peak tariffs are logical and necessary to ensure fuller utilization of the capital investment in generation and transmission plant. This leads us to two further defects of centrally generated electricity as an energy source. The first is that electricity cannot be easily stored. This means that it must be generated as it is used. Although the amount of electricity used varies widely during the day and from season to season, enough plant must be installed to meet the maximum expected load. When the load is less than the maximum, this plant is idle or only working at part of its capacity. The second is that large generating plants are not very reliable, and the larger they get, the less reliable they seem to become. This is particularly true of nuclear plant. In May 1976, for instance, fourteen of the fifty-two US power reactors listed in the *Nuclear Engineering International* monthly summary of operating statistics produced no electricity at all. To produce a reliable system, many reactors have to be interconnected by transmission lines, with enough spare plant installed to meet the maximum expected demand even with some of the reactors shut down. Such a system is so expensive that there is little chance of it ever being an economic proposition for a poor country. Nuclear power and centralized electricity generation are definitely not the answer to the energy needs of less developed countries.

Since so much plant must be installed to make an electricity system reliable, it then becomes attractive to sell electricity very cheaply at times of low demand to keep the plant operating. This off-peak electricity is used predominantly for low-temperature heating, in competition with gas, oil and solar energy. Because solar energy would replace off-peak electricity for low-temperature heating, some spokesmen for the electricity industry claim the introduction of solar energy will actually increase the cost of electricity.

The sun does not shine every day, so that unless a solar heating system has reserve capacity, it will need to be supplemented by some other energy source such as gas or electricity. If electricity is used, sufficient reserve-generating capacity must be installed to be able to boost all the solar heaters in use if there is a run of sunless days. This capacity will not be used on sunny days, so that while the total amount of capital investment required is not reduced by the introduction of solar energy, the sale of electricity is reduced, thereby forcing up the price.

The obvious answer to this problem is to boost solar equipment not with electricity, but with gas, and not to have a reduced tariff for off-peak electricity use, but to have a raised tariff for on-peak electricity use. Gas does not have to be produced or transported at the moment that it is used because the pipe system by which the gas is transported stores a lot of gas and can smooth out the fluctuations in demand. Further, the capital cost required to produce and reticulate gas is less than that for electricity (about $150 as against $1000 in Victoria). When a person buys a new gas stove which uses a kilowatt of power, he commits the public purse to perhaps $100 in extra capital expenditure to provide him with the gas. If he buys a new electric stove, he commits the public purse to an extra $1000, because stoves are used at peak time. The tariff that he pays for electricity used in this stove should reflect this fact. If people using electricity at peak times have to pay more for it, there will be a real incentive to cut down on peak usage and reduce the capital expenditure required for electricity production.

Solar energy is cheaper than electricity for low-temperature heating. Since approximately 25 per cent of all energy used is in the form of low-temperature heat, solar energy can already cheaply provide a quarter of energy needs. At present, low-temperature heating is the only solar technology that has been well enough developed to be widely used and to provide reasonably accurate cost figures. At least one company has been manufacturing solar water heaters in the USA since the 1920s. Solar water heaters have been manufactured in Australia for a long time. Currently the industry is worth about $3 million per year and is growing very rapidly.

The costs of solar energy for higher temperature applications, for the production of liquid and gaseous fuels and for the production of electricity, are still a matter of conjecture, as large-scale plants for these purposes have not been built. It seems likely, however, that solar energy will be cheaper than nuclear energy in all cases, although it will be more expensive than the direct use of fossil fuels in most applications. When the environmental advantages of solar energy are taken into account, and the fact that expenditure on solar energy is contributing towards a permanent, rather than a temporary solution to the world's energy problems, its economic advantages over nuclear power become more obvious.

The 'official strategy' based on nuclear energy will not work. This strategy, as reported by the Flowers Commission, is based on a predicted growth in energy consumption of 5 per cent per annum. Since fossil fuels will not be able to keep up with this growth, it is predicted that an energy 'gap' will develop between supply and demand. Nuclear energy is supposed to fill this gap. The Flowers Commission found that half the probable gap might be filled if 3500 gigawatts (GW) of nuclear plant were in operation by the year 2000, about fifty times as much plant as is currently in existence. This would require the completion, on an average, of three large, 1000-MW, nuclear power stations every week for the rest of the century. The most recent (1976) estimates of how much nuclear plant will actually be in existence range from 1700 to 2250 GW, enough to fill a quarter of the predicted gap. It is notable that as time goes by the estimates of how much nuclear plant will be in existence in the year 2000 have consistently fallen. In 1974 the Organization for Economic Co-operation and Development (OECD) predicted a maximum of 4100 GW of nuclear plant in the year 2000. A year later the same agency reduced its prediction to 2480 GW, not much more than half the prediction made the year before. The amount of nuclear plant currently in existence is only a little more than half what was being predicted in 1973. Obviously, nuclear energy is going to fall far short of filling the predicted energy gap.

The reason has already been discussed – rising costs. The reason for the rising costs is in the long run the increasing political pressure being brought upon the nuclear energy industry to build safe plants and to show that they are safe. The time required to investigate the design of a reactor and declare it safe, to hear objections to the siting of the reactor, and to build in added safety measures has doubled the construction time of reactors. The cost of safety plant and interest and inflation during the construction period have been the main contributors to the cost increase.

Examining Alternatives

If nuclear energy is not going to fill the gap, what can? The immediate answer appears to be: nothing. Solar energy is unlikely to be able to provide us with enough power to fill the predicted gap in time, and there is no other source of energy with even the remotest chance of being sufficiently developed in the time. The only available energy strategy is to reduce consumption, divert large resources to the development of solar energy, explore the efficiency of other renewable sources and gradually phase out the nuclear energy programme at least until such time as a demonstrably safe and socially acceptable reactor and fuel cycle can be developed, an apparently impossible task.

Waste Not, Want Not

A tremendous amount of energy is wasted. This is not surprising. Most of the world's energy is marketed by private or semi-government corporations whose main concern is to operate profitably. Since competition, government anti-inflation policies and other constraints limit the price that can be charged per unit of energy, their strategy is to sell as much as possible. On the part of those who provide it, there is a worldwide vested interest in increasing, rather than decreasing, the consumption of energy.

One of the most obvious examples of where energy could be saved is in private transport. The average medium-sized Australian-made car does about seven kilometres to the litre of petrol (i.e. twenty miles per gallon). Volkswagen and other manufacturers are about to release diesel cars which are claimed to do about twenty-five kilometres to the litre (i.e., seventy miles per gallon), a more than threefold saving. Since the passenger capacity of these smaller cars is almost the same as large ones, and cars are very rarely filled with passengers anyway, the introduction of such vehicles will cause a negligible reduction in the 'standard of living' while making fuel supplies go more than three times further. Since about 20 per cent of our total energy is used by petrol-driven vehicles (cars and light trucks), the use of such efficient cars represents a potential saving of about 10 per cent in overall energy consumption.

It has already been pointed out that electricity is a wasteful form of energy, since about three quarters of the fossil energy used to generate it is lost as waste heat. The use of electricity should therefore be restricted to processes where it is essential. Where cheap gas is available, there is rarely any reason for using electricity for heating. One of the most glaring examples of energy wastage through the use of electricity rather than gas is to be found in Victoria as a result of the competition between the Electricity Commission and the Gas Corporation, both state instrumentalities. As a result of this competition many stoves, ovens, water heaters and items of industrial heating equipment are powered by electricity rather than gas. As the ultimate absurdity, the Electricity Commission plans to build a power station fired by natural gas in the Melbourne metropolitan area, to provide the electric power to operate just those items of equipment that could be running on gas. Three times as much gas will be needed to operate the power station as would be needed to operate the appliances themselves directly.

A full catalogue of current energy wastage would be a long list indeed. Amory Lovins, in his article on energy strategy, previously mentioned, quotes studies which suggest that in the long term the efficiency of energy use in the USA could be increased by a factor of at least three or four. He cites a recent review of specific practical measures which argues that the efficiency of energy use could be doubled through technical changes that could be put into operation by the turn of the century.

Will these things happen? Official energy policy seems to be essentially passive in most countries. Indeed, in Australia it does not appear to exist. The Fox Inquiry found it necessary to recommend, '12. A national energy policy should be developed and reviewed regularly.' (p. 186)

Often the so-called policy-makers are nothing but forecasters, looking at past trends and extrapolating them into the future. The fact remains that an energy gap is predicted. Nothing is available to fill it, so energy shortages will develop. Any country that has not developed a policy of energy conservation to face these shortages will suffer.

Indeed Australians are already suffering, if indirectly. When the Arabs raised the price of oil they set off a worldwide economic recession accompanied by severe inflation. The obvious cure for this recession is a deliberate attempt to increase the overall value obtained for the energy dollar. It appears that the rulers of our country have not considered seriously enough the relationship between energy prices and economic recession, let alone set out to implement policies to utilize energy more efficiently. Energy is such a basic commodity in the economy that it appears to be invisible to many people.

Besides the manifest desire of energy producers to sell as much energy as possible, the barriers to energy conservation are many. Both the manufacturers and users of motor cars seem to prefer petrol-gulping monsters to smaller, more efficient cars. In Australia we have a grossly inefficient, overseas-controlled motor industry protected by huge tariff barriers that lock the country in the dark ages of automotive technology.

Building design and construction are controlled by codes that have little reference to the energy-consuming aspects of buildings. A recent study by the American Institute of Architects concludes that, by 1990, improved design of new buildings and modification of old ones could save *one third* of total current US energy use. Studies in Australia suggest that proper design could enable houses in most parts of Australia to be maintained at comfortable temperatures with no artificial heating or cooling at all.

The capital expenditures required to implement energy-saving technical changes are generally far lower than the expenditure needed to develop new energy sources. Lovins estimates that in America the cost of saving a kilowatt of power may range from zero or even negative in new buildings up to a maximum of about $40, with the most expensive technical modifications costing about $300 per kilowatt saved. This is still very much cheaper than installing a kilowatt of new generation and distribution capacity in the electricity system.

More efficient use of energy will delay the development of the energy gap; it will enable existing energy resources and production facilities to cope with energy needs in the developed countries without shortages and crises. But energy conservation is only part of the answer to the world's energy problems. Fossil energy reserves remain finite: more efficient use will prolong their life but not extend it indefinitely. Conservation will reduce pollution from energy sources, but not eliminate it. Conservation will help the developed countries overcome their energy problems, but it will not necessarily help the less developed countries.

Present world energy consumption is the equivalent of 1.1 thousand million tonnes of coal every year. Of this, the developed western countries, with about 15 per cent of the population, consume 58 per cent. Eastern Europe and the USSR, with about 30 per cent of the world population, use 29 per cent. The remaining 55 per cent of the world population have to make do with 13 per cent of its energy. For every unit of energy used by a person in a poor country, the average person in Eastern Europe or the USSR uses four units and the average inhabitant of the developed countries uses sixteen units. The extremes are even more telling: in 1971 the average citizen of the USA used about fifty times as much energy as the average Indian. Of course, all these figures apply to technological fuels such as coal, oil and electricity and not to food. A satisfactory long-term energy policy for the world must overcome the three problems of limited reserves, pollution and unequal distribution.

Solar Power and Soft Technology

Solar energy can provide the answer. It is a permanent and almost limitless energy resource, its potential for pollution is far less than other energy sources, and it is already distributed over the face of the earth – those countries which have less fossil energy generally receiving more solar energy than the developed countries. Australia is the only developed country with a large area within 30° of the equator, where most solar energy is concentrated. The conversion to solar energy will take some time – perhaps about a hundred years, and it will not simply be a matter of replacing existing energy plant with solar plant. If solar energy is to live up to its technological, social, ecological and political promise, a radical revision of our approach to technology and the world is required.

All natural ecological cycles are solar powered. Solar energy captured by photosynthesis provides all the energy for growth, reproduction and movement in the living world. For a while the stored solar energy in fossil fuels has enabled the industrialized world to forget this. As fossil fuels dwindle the need becomes apparent to integrate energy and material cycles with the natural energy and material cycles. There is not room in the world for two different energy systems – one natural, the other in conflict with nature.

Accepting the general principle that human technology must integrate itself into the natural system, a few general characteristics of the resulting energy technology can be stated, knowing that it is based on the sun. This new approach to technology has come to be called soft technology because it is based on understanding of natural systems, flexibility and human values rather than being an instrument of the conquest of nature and of people.

Amory Lovins, again in his article quoted above, has listed some of the attributes of soft technologies:

They rely on renewable energy flows that are always there whether we use them or not, such as sun and wind and vegetation: on energy income, not on depletable energy capital.

They are diverse, so that energy supply is an aggregate of very many individually modest contributions, each designed for maximum effectiveness in particular circumstances.

They are flexible and relatively low-technology – which does not mean unsophisticated, but rather easy to understand and use without esoteric skills, accessible rather than arcane.

They are matched in *scale* and in geographic distribution to end-use needs, taking advantage of the free distribution of most natural energy flows.

They are matched in *energy quality* to end use needs . . .

Diversity is the key to security, survival and the efficient use of resources. One of the chief features of society and technology is its 'monocultural' nature. More than half the world's food supply comes from four species of plant: wheat, rice, maize and potato. A disease that wiped out just one of these species would leave us in serious trouble. Once man relied for sustenance on hundreds of species. In the interests of efficient wood production, diverse natural forests are replaced by monocultures of one species or another. In the interests of efficient administration, humans reduce themselves to numbers on computer tape. Large centralized energy networks are very vulnerable. A breakdown in just one refinery was claimed to be the cause of a recent petrol shortage in Sydney. A centralized electricity-generating system can be disabled by quite small faults in its control systems. By putting all our eggs in one basket, we lay ourselves open to disaster. An energy system where every house and factory collected its own energy off its own roof would be far more secure. The occasional breakdowns in one or another piece of equipment might inconvenience a few people, but not precipitate a citywide, nationwide or worldwide crisis.

Diversity increases freedom by allowing everyone to have some control over his or her own energy supplies, rather than being beholden to an immensely powerful oil company or electric utility. Diversity allows each energy supply to be tailored to the needs, desires and resources of the consumer. Diversity gives scope for human creativity to develop new means of dealing with problems rather than being tied down by the entrenched practices, rules, regulations and terms of supply of energy authorities.

Diversity provides a basis for flexibility. Flexibility allows for rapid adaptation to changing social and technological needs. A centralized electricity network requires decades and billions of dollars to change. Its very size and inflexibility constrains society to take its energy in the form of electricity, for fear that investment in other forms of energy might be duplication. Nuclear power is the ultimate in inflexible technology. The long-term wastes being generated now are condemning future societies to be organized in a manner that can allow permanent surveillance of these wastes. An energy technology that is based on small, easily-manufactured units will be able to evolve rapidly to take advantage of recent developments and needs. Adaptability makes for strength and durability in the face of a changing world. Rigidity leads to obsolescence and death. This is the difference between soft and hard.

The difference between high and low technology lies partly in the complexity and difficulty of manufacture of each unit of hardware, which is roughly reflected in cost. A typical nuclear installation costs about one billion dollars. A typical domestic system, capable of making a home independent of reticulated energy, water and waste disposal, might eventually cost about $10 000, one hundred-thousandth of the cost of a big reactor. Less complex systems require less investment of expensive human skills. They are better adapted to self-help by individuals, and to the needs of underdeveloped countries. The adoption of such systems will free much of our intellectual effort from the losing battle to make incredibly complex systems such as reactors work safely, so that the effort can be turned to more humanitarian matters such as food and health and social welfare.

Current technology is based on centralized networks distributing electricity, oil or gas. Most of the investment and expense incurred in providing energy lies in these networks. Solar energy is already distributed, so that much of the centralization can be dispensed with and each energy-user can become more or less autonomous. In most parts of Australia, for instance, a well-designed house would need no artificial heating or cooling and its water could be heated by the sun. It is expected that the price of silicon photovoltaic cells, which convert radiant energy to power, will fall to a fraction of present levels in the next few decades. To facilitate storage, electricity produced from such cells could then be used to electrolyse water to hydrogen, to be stored on the premises. This hydrogen could be used for lighting, cooking and as a booster for the water-heating system on sunless days. Further, it might be used to power the family car. Sufficient water can be collected on the roof of the average suburban house to provide for all real needs, and human and other wastes could be composted to fertilize the garden, which could produce a significant proportion of the food required by the people in the house. A system such as this would make it unnecessary for the house to be connected to electricity, gas, oil supplies, water or sewerage systems and there is no reason why such a system should not be more reliable than the present system of large reticulation networks.

Such a technological energy system would begin to resemble the natural energy systems in which myriad species of plants and animals, each well adapted to its particular ecological niche, share energy and materials through complex interdependent cycles. Each plant gets its energy from the sun without having to be wired into a central system. The natural flows of wind and water move immense quantities of material and energy around the biosphere without the need for anything analogous to the large networks of pipes, wires, rails and roads that absorb so much productive effort in our society and contribute very significantly to environmental pollution through oil spills, exhausts and destruction of natural vegetation.

Finally, soft technologies are matched in energy quality to end-use needs. Energy is the ability to do work. The quality of energy is the proportion of it that can be converted to mechanical work – that is, the motion of some mechanical object such as a wheel or a lever. Theoretically all the energy in an electric current can be changed into work: electricity is high-quality energy – it has a quality of one. About half the energy in high-pressure steam can be changed into work – it has a quality of about one half. Only one or two per cent of the energy in lukewarm water can be changed into energy – low-temperature heat is very low-quality energy. Energy itself is neither created nor destroyed: the quantity of it in the universe is believed to remain constant. What does change is the form and quality of the energy. A very important law of nature is that the quality of energy always tends to decrease. This means that less and less can be changed into useful work and eventually the energy becomes useless. Energy arriving from the sun, for instance, has very high quality. When it is absorbed by warm soil its quality becomes very low. The cost of solar energy is roughly proportional to its quality. Low-temperature heat from the sun is low-quality energy and is cheap. Electricity from the sun is high-quality energy and is more expensive. An all-electric economy, based on nuclear power, must use high-quality electrical energy for all purposes, such as space heating, which really only need very low-quality energy. A solar system, on the other hand, can be tailored to provide the quality of energy needed, thereby saving money and resources.

The possibilities of soft technology are endless. Little is known about them as yet, because little effort has been devoted to learning about them. This must be done if our technological society, having exhausted fossil energy, is to be maintained without being obliged to turn to a centralized nuclear power system with all its hazards.

A Nuclear-Free Future

This brings us back to the questions of democracy and strategy which began this chapter. Can we avoid a nuclear future? The answer appears to be yes, if we want to.

The world is held together by communication. All the processes of nature are an endlessly complex communication of information and energy and materials. The processes of society and of culture are communication processes: politics, education, commerce, science, technology – all rely on communication. We have been able to develop theories of communication – theories that tell us what is necessary for good communication, how communication can be improved, and even how it can be stopped. The processes of government are processes of communication. Oppression is above all the denial of information, of education. It depends on closing lines of communication, censoring the media, silencing those who contradict and preventing them from having access to radio, television and printed media. From these theories it can be inferred that society must become democratic and technology must become soft if we are to survive. Large centralized systems – be they power distribution networks, bureaucracies or centrally ruled nations are essentially unstable.

The reason is not hard to understand. Ross Ashby, one of the pioneers of the theory of communication, proposed what is known as the 'law of requisite variety'. The variety of any system or person or thing is the number of different things it can do. A traffic light has low variety – it can shine red, orange or green and nothing more. An animal has greater variety but is limited by behaviour patterns and specific adaptations – usually fish cannot fly nor dogs climb trees. Humans have the greatest variety of all creatures and are able to invent and practise new forms of thought and behaviour. Creativity increases our variety.

The law of requisite variety says that if one system is to control another, the controlling system must have greater variety than the system controlled. This is well illustrated by the problems of minding children. It is a fulltime job for one adult to attend to the needs and desires of one child, to watch and assist and protect him in everything he or she wants to do. It is impossible to do this for two children unless they are restricted in some way, such as being confined to a room or playground, or told not to do certain things. An adult has sufficient variety to control one active child, but not two. The only way to control two children is to reduce their variety by physical confinement or rules of some sort or another.

A person is completely free when his or her variety is unrestricted in any way by physical or psychological constraints. Obviously, we are not completely free: social convention, ownership of property and thousands of other restrictions such as laws and habits restrict the variety of human behaviour. These restrictions make a population governable. A few people in power can only control a large number of people if their variety is restricted by laws, social conventions and dependence on the social system for survival. As people become freer and more independent, the variety of the population becomes greater and those who wish to control society must either increase their own variety or reduce the variety of the population by further laws and repressive measures if they wish to cope.

Ultimately, if people want to be relatively free, like a child being looked after by one adult, the number of controllers must be equal to the number of people controlled. But who would control the controllers? Unless we wish to get into some absurd progression of controllers controlling controllers, we must give up the notion of control by any centralized power and accept that people must become self-controlling if they want to be free. This is true democracy. Any sort of centralized control must deny people some freedom. Only when people become responsible for their own lives can they be free from constraints imposed from outside.

Technology is our means of survival. As long as technology is centrally controlled, people are constrained and their freedom and variety are reduced. Soft technology holds the key to increased freedom. On the one hand, it can be implemented by small groups of people, giving them the means to be independent of the centralized energy technologies. On the other hand, because soft technology is diverse and has high variety, the energy monopolies will eventually reach the point where they can no longer control it because they cannot increase their own variety enough. If individuals were to get energy directly from the sun, water from the rain, recycle wastes and grow their own food, they would no longer be beholden to company boards, governments and unions who may use the consumer as a pawn in their battles for power. This has long been an ideal. Like so many other social developments, it is dependent upon a certain technological base. The print and electronic media, for instance, have made modern liberation movements possible where they were impossible before. The development of soft technologies will form the necessary technological base for a radical increase in human freedom from domination by centralized powers of all sorts.

The rate of development of soft technology will depend on the amount of human effort devoted to it. It will come quickly if people with scientific, technical and other skills devote themselves to its rapid development. It will come slowly if they do not. There is a crisis in the morale of socially aware scientists and technicians. They have seen too many of the developments that held the promise of better human existence turn sour. It is time for them to see that technology must take a new direction – must become the servant of people rather than money – and start to build a new world, a non-nuclear future.

7 The Politics of the Nuclear Industry
Herb Fenn

How It All Started

The men who built the bomb were brilliant. Awesome in their knowledge, they were the scientific giants of the second world war who had unlocked the power of the atom and delivered to the politicians a weapon that could destroy a city at a single blow. No other scientific advent in history could compare with the sheer dramatic impact of 'the Bomb'.

Emerging from the obscurity of wartime secrecy, its creators burst upon the public scene with their reputations made. Scientists who had worked on the Manhattan Project, who had built the bomb and kept the world 'safe for democracy', were held in awe by the public and politicians alike. The world suddenly found itself standing in the hot glare of the atomic age, beholding the creators of a destructive force beyond comprehension. Small wonder that the creators of the atomic bomb had such influence in those early days. And that influence was used.

After the second world war the major world powers embarked on an arms race to develop these new weapons of destruction. 'Power politics' was the name of the game and international pressure forced both East and West to engage in a contest to produce the nuclear weapons upon which their power depended. Teams of scientists on both sides of the globe worked on and perfected the hydrogen bomb, then worked to produce the weapons in mass numbers, along with the sophisticated systems to deliver them to their targets.

Though they worked for different governments with different ideologies, scientists on both sides had one thing in common: they all justified what they were doing by pointing to the advances, supposed or otherwise, being made by the other side. In this way they kept their work going, kept the government money flowing into their field of interest and assured for themselves a respectable place in society. Whether this was done consciously or unconsciously does not matter. An unfortunate characteristic of human beings is that, when they are engaged in a project that is mentally stimulating, financially rewarding and carrying with it high status and personal power, then they are unlikely to be objective about what they are doing or ever question their own motives and social responsibility. As a group they are more likely to formulate reasons why society needs their labour and should continue to support it.

Scientists from the Manhattan Project were no less human than this and, with the personal power they had gained from their wartime achievement, were able to exert considerable influence over politicians and the public alike and convince them of the need for more and better weapons of mass destruction. Along with the military, which had a special interest in new weaponry and considerable political influence, they were an irresistible force in post-war policy making. Only a few scientists expressed concern about the course they were following. Their wartime project complete, most found it easier to continue with their work and not be troubled with matters of conscience. Instead, they turned towards finding new uses for the power of the atom that they had unlocked.

Nuclear physicists had realized from the beginning that electricity might be produced commercially from power plants utilizing a controlled fission process. The principle of these nuclear plants would be the same as coal- or oil-fired stations except that the heat would be supplied by the controlled splitting of uranium-235 atoms. The amount of energy that could be obtained by this process was considerable, so nuclear power began to be discussed as the energy source of the future. At least, it was promoted as such by interested scientists who may have had qualms about the destructive nature of nuclear research up to that time. As a result, nuclear energy became a matter of interest in government circles to politicians who were influenced in no small part by nuclear enthusiasts such as Edward Teller, the father of the hydrogen bomb. Anxious to show their colleagues their command of the new, awesome technology, some politicians began to push for the development of nuclear power. As a result of their efforts, the concept of the 'peaceful atom' was born with President Eisenhower's 'Atoms for Peace' proposal in 1953.

However, nuclear-generated electricity was not seen as commercially viable in the 1950s because of the low price of oil and coal. So development of the 'peaceful atom' became another project of the government's expanding nuclear bureaucracy – another field, after the construction of massive numbers of weapons, that could happily occupy its scientists and experts and also expand their influence. Because of the vast organizations that had grown up around nuclear technology, the pressure to push on into other fields was practically irresistible, particularly when the decisions about financing this nuclear bureaucracy were being made by politicians who were in awe of the scientists who ran it. These politicians did not have the technical competence to challenge what the scientists and experts told them. Thus we were committed to the so-called 'peaceful atom' by politicians whose enthusiasm for nuclear technology was matched only by their ignorance of its implications.

Much the same process worked in Australia, with Philip Baxter in the role of the scientist-lobbyist. As Chairman of the Australian Atomic Energy Commission (AAEC), Baxter played upon the politicians' fascination with science and became a formidable spokesman for nuclear energy in Australia. He and the AAEC managed to get the government committed to the project of building a nuclear reactor at Jervis Bay, NSW, although Australia, with its vast coal reserves, had no need of nuclear electricity. Fortunately, the project was abandoned by the McMahon government as being ill-conceived and uneconomic.

The Rise of the 'Peaceful' Nuclear Industry

The 'experts' had convinced the government to set up and subsidize research into the generation of electricity by nuclear means by convincing them that this was going to be the energy source of the future. Nuclear research went forward in many countries with the government picking up the tab and justifying it with glowing predictions about the future of nuclear power. In the 1950s the boast that nuclear electricity would 'be so cheap we won't need meters on our houses any more' was frequently heard. To the public the promise of a 'peaceful atom' added a ray of hope in a world entangled in a massive arms race. The promise was comforting and by and large the public supported the efforts to find a 'peaceful' atom.

Government subsidies encouraged a wide variety of corporations to devote expertise to developing nuclear power and, as research continued, a nuclear industry was born. Like the scientific establishment that worked on the Manhattan Project, this new industry also had an interest in its own survival and expansion. It was aided in part by such powerful spokesmen as Edward Teller and Hans Bethe, former Manhattan Project scientists. During the 1950s, nuclear power reactor construction began in the USA, the UK, France, Canada and the USSR. These early projects were justified on the basis that, even if they were expensive, it was necessary to gain operating experience with nuclear power in preparation for the day when it was the major energy source. The scientist-lobbyists had had their way and the reactors were off and running. The system that various governments had set up with their advice was ideal for the infant industry. Public money funded most nuclear research and the industry was 'regulated' by sympathetic government agencies.

In both the USA and Australia this government body was called the Atomic Energy Commission (AEC) and it was charged with both regulating and promoting the industry. It was not until some years later that these two functions were seen to be in conflict. In an effort to eliminate this conflict of interest between regulation and promotion, the US Atomic Energy Commission was split into two agencies in 1975: the Nuclear Regulatory Commission and the Energy Research and Development Administration. This has yet to happen in Australia and the Commission here continues to emphasize its promotional aspect. The AAEC has the function of 'regulating' the industry by the laying down of rules and procedures and of 'promoting' the industry by encouraging exploration for uranium, educating the public and doing research. It was supposed to be an impartial body dealing in scientific facts that would help bring the world into the nuclear age. However, it soon became evident that the AAEC had a vested interest in seeing the growth of atomic energy because, as the industry grew, so would the government body that regulated it. As a result the 'promotional' aspect of the AAEC's responsibilities assumed the greatest significance and made a mockery of any claims of impartiality on the part of the Commission. It wanted to see nuclear power grow and interpreted 'scientific facts' in that light. Hence, the

Commission pushed the idea of the reactor at Jervis Bay and was nearly successful in having it built.

After the Jervis Bay Scheme was scrapped by the McMahon government, the morale at the Commission sank dramatically as a 'sense of mission' had been lost. The AAEC now admits that the construction of nuclear reactors here in Australia cannot be justified before 1990 and has turned to other areas in which to promote nuclear energy and thus justify its existence.

The most obvious area for this has been in the uranium mining question where the AAEC has been a leading advocate for mining and export. This has been particularly evident in the Fox Inquiry which has considered all aspects of nuclear power. By its attitude during the cross-examination of witnesses opposing nuclear power, the AAEC revealed its determination to promote mining. It was evident at the Inquiry that the mining companies and the AAEC were on the same side of the table. In its testimony to the Inquiry and in giving advice to all levels of government and the public, the AAEC tried to put nuclear power in the best possible light. Its projections of growth in the world-wide industry are always optimistic and any hazards associated with nuclear power are minimized. The AAEC seems to be the victim of excessive positive thinking and, as a result, is afflicted with a severe case of tunnel vision.

For instance, spokesmen for the AAEC, when confronted with a question of the toxicity of plutonium, automatically counter it with a reference to *Botulinus* toxin (sometimes found in preserved food), saying it is the more dangerous so we should not worry about plutonium. Their argument is as weak as it is irrelevant. Just because *Botulinus* is dangerous does not mean plutonium is safe, particularly since the latter can be made into nuclear weapons and will be produced in large quantities if the nuclear industry goes on as planned.

Why does the AAEC have a bias towards the case for uranium mining and nuclear power? Simply because it knows that if uranium mining is stopped, then the Commission has no future. It is unlikely that, if Australia turns away from nuclear power by leaving its uranium in the ground, it will ever commit itself to building nuclear power stations and expanding the nuclear industry here. Thus the AAEC will be relegated to doing nothing more glamorous than producing medical isotopes at Lucas Heights.

The Nuclear Establishment

Nuclear power is big business and requires a vast input of money and expertise. For this reason the companies who originally showed the most interest in this technology were those with a large organization and reserves of capital. Nuclear power fitted in very nicely with their already established organization which had been built up to harness complex technologies. Any reservations they had were eliminated by the abiding faith in the future of nuclear power evident in the 1950s, the possibility of huge profits or the influx of government money that encouraged nuclear research. Billions were invested in the future of nuclear power by both governments and private corporations. There was little or no questioning as to the desirability of nuclear-generated electricity. It was going to be 'safe, clean and efficient'. The same competent scientists who developed the atomic bomb said so. And few people dared question them.

The new energy technology was of more than passing interest to the world's oil companies who were well aware that the oil they controlled would one day be exhausted. Companies such as Standard Oil (NJ), Gulf and Kerr-McGee began to invest in the various fuel aspects of nuclear power, uranium mining, fuel fabrication and fuel reprocessing. The investments they made were enormous and as they grew so did the companies' vested interest in the future of nuclear electricity. Leading financial institutions such as the Chase Manhattan Bank also invested heavily in this capital-intensive enterprise as did some of the world's richest and most influential families, such as the Rothschilds and Rockefellers. Therefore, some of the largest corporations, financial institutions and influential families acquired a growing interest in seeing nuclear power as the energy source of the future. This 'nuclear establishment' now wields great political power which has helped the industry overcome conventional barriers that might otherwise have stopped it.

One of the first problems encountered by the industry in the USA was the question of insurance. Who would be liable for the thousands of lives lost and billions of dollars in property damage if a catastrophic accident occurred in a reactor? Private insurers were not willing to take such a risk. The US government came to the rescue in 1957 with the Price-Anderson Act which limited a utility to responsibility of damages of $560 million if an accident occurred. At the time the damages that would occur from a large nuclear accident were estimated to be seven billion dollars. Present-day estimates of property damages in a nuclear accident run as high as fourteen billion dollars but, due to the powerful political influences of the nuclear establishment, the Price-Anderson Act has not been changed.

The massive power of the nuclear energy lobby could be clearly seen after the 1973 oil crisis. Following the reduction in oil supply from the Middle East, the Nixon administration formulated 'Project Independence', a scheme whereby the USA would become independent of foreign energy supplies by a massive expansion of its nuclear programme. The proposal ignored the possibility of developing a wide range of alternative energy sources and settled for a single-minded commitment to breeder reactors. The public was led to believe that there was no alternative.

The funds available to the nuclear establishment to press its case with the public were decisive in the US referenda on nuclear power in 1976. In Washington State alone the pro-nuclear forces outspent the opposition by seven to one. The money bought access to the media and it was this media promotion that won the referendum. A similar media campaign has been launched in Australia as the mining companies meet opposition from the public.

With its political muscle, the industry and its supporters have achieved a level of public funding for nuclear power that far surpasses the spending on any other energy source. For every dollar that is spent on solar energy in the USA, eight are spent on one nuclear project alone, the breeder reactor.

Much the same situation exists in Australia with nuclear research getting the major share of the government's energy budget. The lingering influence of Sir Philip Baxter and Sir Ernest Titterton, Australia's leading nuclear proponents, has helped ensure that the AAEC has been well funded, though the organization has little to show for twenty-three years of effort except a hole in the ground at Jervis Bay.

The nuclear establishment in Australia is composed of the mining companies, organized as the Australian Uranium Producers Forum, and the AAEC. They are supported by Baxter and Titterton who act as the leading 'nuclear hawks', along with a few attendant academics. Rounding out the cast are some outspoken individuals such as the Premier of Queensland Joh Bjelke-Petersen and mining magnate

Lang Hancock, both of whom have varying interests in uranium mining and are not above equating Australia's interest with their own. With their financial and political influence, such people are seeking to push Australia down the nuclear path with as little discussion as possible.

Lang Hancock has long been an advocate of using nuclear explosives to mine iron ore and create harbours from which it can be shipped. Hancock, who makes a fortune from his Pilbara iron mines, feels that nuclear explosives could bring down the cost of iron mining substantially. His advocacy of uranium mining seems to be based on the idea that it is the first step towards using nuclear explosives in iron mining. If the Australian public will accept uranium mining, then it might also accept his proposal and he would be richer as a result.

The mining companies have kept a fairly low profile until recently when they realized that the public image of nuclear power was not the best. In an effort to correct that and 'sell' the idea of nuclear power to the public, they imported a film from the USA and made it freely available to all schools in Australia. The film, entitled *Now That the Dinosaurs are Gone*, is a slick propaganda production that avoids the major issues in the nuclear debate. There is no discussion of nuclear weapons proliferation, terrorist diversion of plutonium, or the long-term storage of radioactive waste. The film shows only clean, white reactors and green landscape, in an effort to convince the viewer that all is well with the nuclear industry. Because the mining companies stand to lose millions if uranium mining does not go ahead, they can afford to spend money on the distribution of such a film. They can also afford to print expensive leaflets on the supposed benefits of their industry and, because the opposing viewpoint does not have the monetary resources to counter them, they can be successful in convincing at least some of the public. The idea of 'getting the message to the kids' is an old one used often by people with something to sell. Children are more impressionable than adults, yet can influence their parents by what is learnt in school; so one way to convince parents that uranium mining is safe is first to convince their children.

The psychology used by the nuclear industry in putting its case also takes advantage of other weaknesses in society and human nature. People tend to be dazzled by the flashing lights of complicated technologies and are taught to respect the 'learned' men and women who devise and handle them. Society as a whole finds it easier to take their advice than to question it. You must 'trust the experts' is a phrase often used by nuclear proponents to justify their actions and to convince the public that the questions involved in the nuclear debate are beyond their capacity to understand.

However, as the First Fox Report noted, the primary questions are not technical in nature and should not be left to the experts:

Ultimately, when the matters of fact are resolved, many of the questions which arise are social and ethical ones. We agree strongly with the view, repeatedly put to us by opponents of nuclear development, that, given a sufficient understanding of the science and technology involved, the final decisions should rest with the ordinary man and not be regarded as the preserve of any group of scientists or experts, however distinguished. (p. 6)

Equating the advance of nuclear power with 'progress' is one of the favourite emotional appeals of the industry. Society has been imprinted with the idea that 'progress' is by definition good and desirable and that anyone who questions it is a bit odd and possibly a Communist. The logical extension of this idea is that, if we don't have 'progress' in the form of nuclear power, then we shall have to go 'back to the caves' to a subsistence form of living. This is, of course, simply propaganda that plays upon existing fears in society. If nuclear power with its far-reaching hazards does represent 'progress', then society had better ask the question 'progress to where?'

Various other arguments are put forward by the industry to try to justify the mining of uranium. Most are poorly thought-out propaganda. One that has been used by the head of the AAEC, Professor Don George, is that we must export our uranium to help the Third World. Helping those less fortunate than ourselves is an admirable motive. However, to apply it to the export of uranium is simply an emotional appeal designed to fool those members of the public who are naive enough to believe it. The truth is that nuclear power stations are being constructed for the industrialized nations of the world who can afford them, not the poor nations whose need for energy is greatest. The capital requirements are simply too great for most nations of the Third World. In any case most of them do not have have an established electrical grid to take the load even if they could afford the technology. Nuclear power is a rich man's energy source so Australia's uranium exports will benefit only those who are already well off. As was pointed out in the First Fox Report, Australia could do a service to the underdeveloped world by researching alternative energy systems that can be adapted to their needs as well as the needs of the industrialized nations. Solar energy systems can be developed on a small scale that would find application in underdeveloped countries, as many of them lie in tropical and subtropical areas.

An argument often put forward by Sir Ernest Titterton is that the world has an insatiable need for energy so we must do all we can to supply it. This is a very simplistic argument and takes a very narrow view of the world's energy situation. There is no doubt the world needs energy, but it is important to note what form the requirements take. The most pressing need for energy is in the form of liquid fuels as the world is fast running out of petroleum. Nuclear reactors only provide energy in the form of electricity, not liquid fuel, so they are not applicable to our most pressing energy needs. Also, electricity is only required for about 15 per cent of our end-use energy needs: for lights, electrical motors and the like. Most of our requirements for energy are in the form of either low- or high-temperature heat. Such requirements might be met by utilizing solar energy without the hazards associated with nuclear power, as described in the previous chapter.

The mining companies often argue that if Australia keeps its uranium in the ground, it will make no difference to the development of nuclear power worldwide as countries such as Japan and the USA will simply buy their uranium elsewhere. This argument ignores the political realities of the world today and the needs of the nuclear industry. The industry, because it needs long lead times for the planning and construction of reactors, must have a source of uranium it can rely on. Also, it needs uranium that is rich and easy to mine. Australia has the largest and richest reserves in the southern hemisphere and is doubly attractive because it has good prospects for long-term political stability. In contrast, southern Africa also has large uranium deposits but its prospects for political stability are poor, to say the least. The nuclear industry does not want to rely on a fuel supply that may be cut off at short notice. Thus, they are looking to Australia as a major uranium supplier. If Australia does not supply, the impact will be great and, with other events, may signal the end of the nuclear industry.

The final and most deliberately emotive argument used by individuals in favour of uranium mining is that if we do not sell it the Japanese will come and take it by force. This argument was put forward by Sir Philip Baxter and the Minister for National Resources, Doug Anthony, in separate statements barely forty-eight hours apart in March 1976. The two statements raised the fear of the 'yellow peril', the spectre of little yellow men from the north sweeping down to take our valuable uranium. That men of supposed high standing in the community would use such scare tactics certainly does not reflect well on them. Japan has contracted for its uranium until 1985 and has no pressing need for Australia to be a supplier. Also, Japan is not the military power it was in 1940 and does not have the army or navy to sustain a war thousands of kilometres from its own territory. Thus, the threat of Japanese invasion is merely a tactic used to delude the Australian public by those with a vested interest in uranium mining.

The Multinational Connection

In the last fifteen to twenty years the world has seen the dramatic growth of organizations described as multinational corporations. With the growth of world trade, technology and expansion of the material standard of living, corporations have expanded their operations into a variety of countries to open up new markets and take advantage of cheap labour and tax loopholes. In the process they have become the controllers of a large share of the world's wealth. In his book, *World Without Borders*, published in 1973, L. R. Brown points out that of a list of the one hundred wealthiest nation-states (based on GNP) and multinational corporations (based on gross annual sales), fifty-nine are nations and forty-one are multinational corporations.

Controlling such economic power brings with it great political influence. This influence over world affairs has grown as the multinationals have grown and, because of its international scope, has been beyond the control of national governments. However, recently, the US government has begun to explore the possibility of anti-trust action against some of these organizations. The legal process is a slow one and results may be a long time coming. Meanwhile, the power of these corporations is a reality and is used to perpetuate and expand the organizations often to the detriment of native populations that do not have the means to resist. The countries that are most liable to feel the pressure of multinationals are those with large reserves of natural resources, such as Australia.

Much to the alarm of ecologists, the industrialized nations of the world are using up the world's non-renewable resources at a rapid pace. The USA, once self-sufficient in nearly everything, has now had to look overseas for the oil and other resources it needs to maintain its growth. Japan is in a worse position as it is a resource-poor country and is trying to maintain its post-war growth rate by huge imports of raw materials, many of which are made into products that are used once and thrown away. Increasingly, it is looking to Australia to help it maintain its 'cowboy economy'.

Thus Australia comes under pressure from both foreign governments and multinational corporations to export its minerals both cheaply and in great quantity. For most of its history, Australia has been content to be a supplier of raw materials. However, it may be no coincidence that in times when governments have chosen not to supply, or to supply at a higher price, they have toppled. Such was the fate of John Gorton's government when he surprised his Liberal colleagues by choosing to husband Australia's reserves of uranium and bauxite. Gorton's scrupulous honesty in refusing shares offered by Comalco in their bauxite venture revealed the sorts of practices that are resorted to in an effort to co-opt politicians. Multinationals control large amounts of wealth and some are not above using it to gain their ends with politicians. The Lockheed scandal was a good example of how far bribery can go. When multinationals, through such techniques as bribery, can influence political decision-making to a far greater extent than can the population at large, the concept of democratic government becomes a farce.

In trying to conserve Australia's resources, Gorton ran up against the interests of Rio Tinto Zinc, the world's largest mining company. RTZ, which is based in the UK and holds 15 per cent of the world's known uranium reserves, retains a controlling share over its local subsidiary, Conzinc Riotinto Australia (CRA). In turn CRA owns controlling shares in several companies, including Mary Kathleen Uranium and Comalco, against whom John Gorton had tried to stand. The

Labor government under Gough Whitlam pursued a minerals policy that kept many Australian minerals in the ground for three years, and also ran up against the mining giants, and also toppled. No democratically elected federal government in Australia, Liberal or Labor, is immune from their power for very long. The result is that the Australian people have little control over their mineral wealth unless they demand that their elected representatives regain control of the situation.

The existence of worldwide cartels controlling the world's resources and making national governments obsolete is a spectre that is not far removed from today's reality. Rio Tinto Zinc was one of the original members of the World Uranium Club, or as it was officially titled, the Uranium Marketing Research Organization. This organization, originally set up in 1972, had five members: Canada, South Africa, France, Australia and Rio Tinto Zinc. The only company represented in 'The Club' was RTZ, but as it had uranium interests in Australia, Canada and South Africa's protectorate of Namibia, its influence was considerable.

The ostensible purpose of 'The Club' was to conduct research and exchange marketing information. However, it has become apparent from recent developments that the primary purpose was to fix prices and divide the market. When the organization was formed in 1972, uranium prices were low, about $US15 per kilogram. The boom in uranium during the 1950s and 1960s had tapered off as the superpowers found they had more than enough nuclear weapons to destroy each other, so the market for uranium was at a low ebb. Holding 15 per cent of the world's known uranium reserves, RTZ was interested in seeing a higher price for uranium to allow development of its Rossing project in Namibia in southern Africa. By raising the price of uranium, RTZ accomplished two things. First, it made the Rossing mine more economically viable; and second, it encouraged the South African government to maintain its hold on Namibia, maintain the status quo there and assure a stable atmosphere in which the RTZ mine could operate. In this way the power to control prices can be translated into political action with far-reaching effects.

Since 1972 the price for uranium has risen $US88 per kilogram and there is evidence that the rise has been due to price fixing by an international cartel that evolved from the Uranium Marketing Research Organization and includes most of the uranium mining companies in Australia. Both the US Justice Department and the Westinghouse Corporation have taken legal action in relation to the cartel, with Westinghouse suing twenty-nine uranium producers for price fixing. However, the Australian government in November 1976 acted to thwart US litigation by passing urgent legislation, the Foreign Proceedings (Prohibition of Certain Evidence) Act. This Act was designed to protect local uranium producers from probes into their business practices and, in effect, gives protection to the Australian members of the price-fixing cartel. The speed with which this legislation was introduced is an indication of the power of the uranium lobby. (At the time of writing there seems to be some question as to whether this hastily passed legislation is constitutional. If it is found to be unconstitutional, it will be interesting to watch what further actions the uranium miners will take to protect themselves. Like Richard Nixon, they have a lot to hide.)

The fact that the uranium cartel has kept prices so high has influenced political thinking in Australia. The pressure to allow mining to take advantage of the high price has been great and has made the job of anti-uranium mining forces much harder because 'money speaks with a loud voice'. The ability of a cartel to raise the price of uranium to an artificially high level can effectively destroy any opposition to mining by making the economic gain appear irresistible. Over the last year the Australian public has heard glowing statements of how our uranium will make us as rich and powerful as the Arabs. The average person is often fooled by arguments such as these, particularly if times are bad economically. What the Australian public has not been generally told is that they will receive little of the money from uranium mining.

One of the interesting findings of the First Fox Report was that even if uranium prices stay high, 'The Ranger project would probably generate a substantial rate of the capital invested. However, its contribution to net national income and employment opportunities would be relatively small.' (p. 83) In other words, most of the income would accrue to large investors in the industry and to the mining companies, many of whom have strong multinational connections. The Australian public would gain little from the sale of its uranium. The taxes paid by the companies into the public purse would hardly offset their debts to it for research subsidies, export incentives, monitoring agencies and even, environmental inquiries.

The extent of multinational influence on the Australian uranium scene can be noted by examining the company Pancontinental Mining. Pancontinental is the holder of the Jabiluka deposit which is the largest uranium deposit in Australia and possibly the world. A US company, Getty Oil, has a 35 per cent interest in the Jabiluka deposit and is committed to putting up considerable finance for the project and providing an assured market for the uranium once it is mined. Pancontinental has also entered other joint ventures for uranium exploration, in both Australia and Canada, helping to cement its ties with North America. With such powerful overseas connections, Pancontinental can wield considerable political influence in the world's uranium markets and with the politicians of several countries.

Often the multinational connections are numerous and complicated, as is the case with Queensland Mines, holder of the rich Nabarlek deposit in Arnhem Land. Fifty per cent of Queensland Mines is owned by Kathleen Investments which is an associate company of Mary Kathleen Uranium Ltd. Mary Kathleen is owned and controlled by Conzinc Riotinto Australia (CRA) which itself is owned by Rio Tinto Zinc of the UK. The second-largest shareholder in Queensland Mines is Noranda Pacific which is a subsidiary of Noranda Mines Ltd of Canada. The fourth largest shareholder is Newmont Pty Ltd which is a subsidiary of Newmont Mining Corporation, USA. Queensland Mines is also an associate company of Imperial Chemical Industries Ltd of the UK.

As these two examples show, foreign interests in Australian uranium are extensive. Foreign capital has been invested both directly and indirectly in prospect of great returns in the future. The stakes are high. Australia's uranium is worth about $25 000 million a year on the world market at today's prices. To protect this investment and assure that mining does go ahead in the future, companies have expended effort to influence the political process. The magnitude and scope of these efforts is unknown but, with so much at stake, it is reasonable to assume they have been very extensive.

The Energy Companies

As mentioned previously, oil companies have shown considerable interest in nuclear power and many have invested heavily in its future. Since the second world war oil companies have sought to diversify their holdings into, not only uranium, but also coal, tar sands, oil shale and natural gas, so that today they are not so much 'oil' companies as 'energy' companies. In 1971, US 'oil' companies milled 40 per cent of US uranium and 20 per cent of the coal. One oil company, Humble, was the second largest coal owner in the nation. Thus, the whole energy industry is tending toward an oligopolistic structure composed of a few giant companies controlling a wide range of energy sources.

To exploit the energy in coal, oil and natural gas requires high technology and the large amounts of money that energy companies can command. Drilling for oil and natural gas on the ocean floor is a difficult business that requires a large organization with much expertise. The energy companies have built up such an organization suited to the technological challenge.

The utilization of nuclear power is also a high technology operation requiring expertise and money. Thus, the expansion of energy companies into this field follows rather easily as they already have the organizational base to build upon. However, not all energy companies are finding the nuclear business an attractive proposition. Gulf General Atomic, a subsidiary of Gulf Oil (US), pulled out of the nuclear business in 1975 because of losses it incurred. Because of the capital requirements and the economic problems the nuclear industry has encountered, only a few companies will be able to remain in the business. For that reason, nuclear power will tend to be an energy source that is more monopolized than oil, coal or natural gas is today.

The companies who monopolize it will be the largest of the large. Already in the field of reactor components, two companies dominate the world market, Westinghouse and General Electric. They control 85 per cent of the nuclear component business. The uranium-mining stage of the nuclear fuel cycle is tending toward a worldwide cartel arrangement which fixes its own prices. The uranium enrichment and fuel reprocessing stages are presently dominated by government enterprise, even in the USA. However, there have been moves to turn these over to private corporations. If this happens, monopolization of these areas of the fuel cycle will also occur, as they require both capital and a large body of expertise.

All phases of nuclear power require high technology and, for this reason, this power source can be developed only by the energy companies who have built their organizations on high technology. Because of its very nature, nuclear energy will be controlled by only a few companies. If the world moves towards a massive dependence on nuclear power for its future energy needs, then it will be handing over a tremendous amount of economic and political power to those who control it. Whether this is a wise course of action must be decided by society before it embarks on massive nuclear development.

Nuclear Politics in Socialist Countries

Due to the lack of hard information regarding the nuclear industry in the USSR and Eastern Europe, many conclusions here can, of necessity, only be inferred from what information is available. The USSR claims to generate about 2 per cent of its energy needs from nuclear power and has exported reactors to Finland and Eastern Europe. A nuclear industry, whether operating in the East or the West, will have certain characteristics in common. First, both require a large well-funded organization to operate effectively. Secondly, both require large numbers of experts with technical knowledge. This situation must exist in the USSR and, to some extent, in China. The nuclear programme in the USSR was an outgrowth of their development of the atomic bomb and the subsequent arms race with the West. As was the case in the USA, a large organization grew up around nuclear weapons production and this organization, like its Western counterpart, had a vested interest in its own survival and expansion. It could justify its expansion by pointing to progress in the West on nuclear power, particularly by raising the fear of a rearmed and economically powerful West Germany. The Soviet Union has paranoid fear of the Germans because of the second world war. The need to be strong both militarily and economically can be easily linked with the idea of nuclear development, whether for bombs or for 'peaceful' purposes, simply by raising memories of the war. The argument would be something like this: nuclear bombs make a nation powerful militarily; nuclear power stations make a nation strong economically by supplying the energy for industry. The second world war showed us the price of weakness; therefore we need the nuclear industry to make us strong and keep us safe. The reader will notice that this argument is similar to the ones offered by some nuclear proponents in the West.

How much political power the nuclear industry has in the Soviet Union is difficult to assess. If Soviet politicians are as easily impressed by large-scale technology as are those in the West, then the political power of the Soviet nuclear industry is considerable. This may also be inferred from reports of nuclear accidents in the Soviet Union. In spite of reported accidents with waste storage and the breeder reactor, the Soviet programme continues to press ahead, though possibly at a reduced level. What opposition there is to it cannot be expressed publicly, as it can be in the West but, even though public opposition cannot be expressed, a combination of several factors may serve to bring the nuclear programme of the socialist countries to a halt:

1) Economic - Costs of nuclear power are rising and will make it less attractive as an energy source.

2) Accidents - The capacity for human error and human irrationality is the same in both East and West. A few more accidents of the nature reported so far will give ample reason to scale down the Soviet nuclear programme.

3) The decline of the nuclear industry in the West - If the West abandons nuclear power because of rising opposition, the pressure on the Soviet Union to keep abreast of Western advances - their strongest motive for a continuing nuclear programme - will be relieved.

However, nuclear power is more easily accommodated in an authoritarian state. It is a highly centralized form of energy that requires heavy security for its safe operation. Shipments of radioactive material, as well as reactors, must be guarded from sabotage and the police must have broad powers of search and seizure to recover stolen nuclear material, as described more fully in chapter 5. Such measures are already the stock in trade of an authoritarian regime, and the difficulties of imposing them would not be encountered there to the same degree as they would in the West.

The Soviet obsession with nuclear power should therefore provide an object lesson for the Western world. To parallel the Soviet controls would mean facing the possibility of the West becoming equally authoritarian and repressive. It is significant that the arguments of the Soviets and those of nuclear proponents in the West are identical.

It is in the interest of both to preserve their elitist, authoritarian position even at the risk to the world's population of the hazards of nuclear power and nuclear war. Fortunately, Australians can still speak out against the erosion of civil liberties and danger to life and limb that nuclear power would undoubtedly cause.

What the Nuclear Establishment Fears

As has been mentioned, the nuclear establishment is composed of some of the world's largest corporations, government agencies with a vested interest in the future of nuclear energy, and powerful individuals. Nuclear energy is the key to their economic and political domination. If it grows to be the world's major energy source, then the power of Westinghouse, General Electric, Standard Oil, Nelson Rockefeller and the various atomic energy agencies around the world will grow accordingly. Energy is basic to society and those who control the world's energy do, to a large degree, control the world. That is the lesson that the Arabs taught the West in 1973. The prospect of controlling the major energy source of the future is certainly a tempting one for those who seek power and money.

'Nuclear leasing' is the latest concept in this field. An international company buys up uranium and leases it to utilities in various countries. Then the leasing company retains control and ownership of the plutonium produced in fission reactors for further use as fuel in breeder reactors. Over a period of years, as the world turns more and more to the breeder reactor, the plutonium held by the leasing company comes into greater demand. With a full-scale commitment to breeder reactors, the few thousand tonnes of plutonium that the company holds becomes, in effect, the world's major energy source. According to documents leaked to Friends of the Earth in November 1976, discussions involving such a leasing plan have been going on between Peko-Wallsend, Nuclear Corporation International, British Nuclear Fuels Limited and the Iranian government.

Once companies and governments are committed in this way to breeder reactors, they will begin to use their political power to assure their investments pay off. Adding to the nuclear juggernaut, the powerful seek yet more power. Because of the power and money it controls, and the momentum it has already attained, the nuclear establishment at first glance appears unstoppable. A closer examination, however, reveals that this is not the case and there are certain trends both in society and in the nuclear industry that the industry greatly fears.

The first of these is the growing public opposition to nuclear development worldwide. There have been occupations of reactor sites in France, Germany and Switzerland, civil disobedience in the USA and union action against uranium mining in Australia. Ralph Nader has called nuclear power a 'technological Vietnam' and made the issue his number one priority. If the opposition continues to grow, the nuclear industry will be in real trouble. Secondly, the economics of nuclear power stations have not been good. In the USA, sales of reactors have plummetted because of high costs and a levelling off in electricity demand, as well as public opposition. As a result, nuclear vendors have tried to find a market overseas and have opened a vigorous campaign to sell power stations in the few Third World countries that can afford them. The reactors are not sold as efficient producers of power so much as national status symbols. The fact that it is only a short step from a peaceful power station to the manufacture of an atomic bomb is also a vital consideration even if it is not stated in the sales contract.

There are trends in Western society that also endanger the nuclear establishment. The first of these is the growing interest in alternative sources of energy. Nuclear power has clearly not lived up to its boast of being safe, clean and efficient, thus more research money is gradually going into solar, wind, geothermal and other energy sources. A characteristic of most of these alternatives is that, unlike nuclear power they are difficult or impossible to monopolize. They can also be operated on a small scale, unlike nuclear power stations which can only operate economically on a large scale. A move by society away from large-scale centralized nuclear power to small-scale, decentralized alternatives endangers the political power of the nuclear establishment by striking at its very roots. A movement by society towards decentralized energy production where each house or neighbourhood supplies its own energy needs with solar collectors, windmills, etc., means the decline of the nuclear establishment and all those organizations with a vested interest in centralized power production.

Seeing this danger, large corporations with interests in a continuation of centralized electricity have devised various plans of their own to harness alternative energy in a centralized manner and thereby maintain control. One such proposal is to build a giant solar collector in space and beam down the energy in the form of microwaves to a central point. From this point the energy would be distributed as electricity. This high-technology utilization of the solar power appeals to large corporations but there are better ways to utilize the sun's energy, as Chapter Six shows.

However, the corporate establishment will continue to push for adoption of such grandiose schemes and they may be aided in their efforts by other vested interests such as the military. A solar collector in space beaming down energy as microwaves could have a rather bizarre military application. Microwaves are known to the public by their association with microwave ovens which cook food in a very short time. If microwaves from a solar collector in space were focused on a city, the effect on the inhabitants would not be pleasant. The military establishment in the USA can be expected to show interest in a scheme such as this and may use their political power in support of it.

The second trend in society that endangers the nuclear establishment is the rise of new styles of living that embrace low consumption, low energy use and more self-sufficiency. To some degree this trend can be seen in the USA where, for the first time this century, the pattern of migration has shifted from city back to rural life. More people are leaving US cities than are moving into them. Not all the migrants are adopting low-energy lifestyles, but the trend is towards greater self-sufficiency and a more harmonious relationship with nature.

If this trend towards decentralization continues, then an increasing emphasis on alternative energy systems can be expected. In a world powered by alternative sources of energy, political power would no longer rest only with the centralized establishment, but in the hands of individuals who, through their self-sufficient lifestyles, have control over their own energy supply and hence, over their own lives.

The nuclear establishment tries to typify this trend as a step backward in the progress of the human race. To those who describe progress only in terms of mass-produced gadgets and complex technology, the response is predictable. Human progress is a matter of definition and in any discussion of that subject the question 'progress to where' should again be raised. In considering the question, society must examine the political implications of allowing a handful of international companies to control the world's major energy source, for that is what the nuclear establishment is trying to do. Their economic and political weight in society today is massive and they are gambling that, because of their sheer size and influence, society will have no choice but to submit to their plans for a nuclear future.

Conclusion:
Stopping the Atomic Juggernaut

The origin of the word 'juggernaut' is the Hindi word *Jagarnath*, which means 'Lord of the world' and was a title of the god Krishna. An idol of the god used to be dragged on an enormous cart in an annual procession through the streets of an Indian town. Disciples of the cult threw themselves under the wheels to be crushed. So, concludes the *Oxford English Dictionary*, the word has come to mean anything to which persons blindly devote themselves, or are ruthlessly sacrificed. It is a fitting description for the nuclear industry.

We have described the sacrifice that nuclear power demands. If mining goes ahead in the Northern Territory it will mean the destruction of part of the world so rich in flora and fauna of a unique kind that it has been classified as part of the world heritage. The Aboriginal traditional owners of the land who have lived in harmony with this heritage for 20 000 years would also be destroyed: they would become, instead of proud inheritors of their culture, labourers for the firm of Ranger and outcasts of a mining town.

'*Balanda* (white man) push, push, push – soon pubs everywhere and they will kill the race. Look at the Larrykeahs; Darwin is their country and they are living on the tip' (Oenpelli tapes). The Aboriginal people of the Northern Territory would be the beginning of the sacrifice on the all-consuming altar of nuclear power. The riches of Australia, which to date belong exclusively to the invading European, have led to the ruthless exploitation of a brave, imaginative, ingenious people. The Oenpelli expect the same pattern to be repeated and that dominant European 'civilization' will, through force of numbers, money and political influence, overwhelm them. The prevention of uranium mining on traditional Aboriginal lands, the protection of sites as sacred to the Aborigines as St Peter's is to Catholics, or the Wailing Wall to the Jews, is the last opportunity Australia will have to exculpate to some degree the shame of the fatal European impact on the original inhabitants of this continent. At a national conference of organizations opposed to uranium mining in December 1976 in Sydney, a representative of the Aboriginal Land Rights Movement thanked the conference for its stand. It was not said then, but it should have been said, that it was a travesty for Aboriginal people to thank Europeans for not mining their land. It compounds the shame, especially for a society that purports to base its laws on respect for individual rights.

At a public meeting in Martin Place, Sydney, a man appeared from the crowd and approached the speakers. He claimed he had mined uranium for years and was still perfectly fit and healthy. 'We get covered in the stuff,' he said, 'but it's just a rock. We don't even wash our hands.' It is not surprising that such an attitude persists among uranium miners and workers in the nuclear industry, for they are told that its operation is safe. This is the criminal result of pro-nuclear propaganda which minimizes the hazards to the health of workers in the industry to the extent that people do not bother to take even the most elementary precautions. Even while operating normally, nuclear power is exposing those connected with it to the health risks described in Chapter One. The long-term effect of genetic decay, causing a steady increase in cancers and mutations, is a real possibility. Even if major reactor accidents could be avoided and nuclear war averted, the end result of nuclear power could be a human race crippled by ever-compounding and irreversible genetic damage. In this way nuclear power could destroy the infinitely complicated and delicate mechanism of the human body, the creation of millions of years of evolution. To watch the remarkable TV series *The Ascent of Man*, by Jacob Bronowski, describing the long, painful, but glorious evolution of all life from its mysterious beginnings, to feel Bronowski's sense of wonder, and then to contemplate the rising level of radioactivity that could destroy it all, is to experience a rising sense of anger that a few men in a few years could risk sacrificing the whole creation of the earth to their passion for the power of the split atom. This is why so many biologists and medical people can only contemplate the spread of nuclear power with horror.

The danger of a major reactor accident is also a real one. Increasing numbers of reactor engineers are beginning to recognize this and reject nuclear propaganda which implies that, although there may be technical problems, these will be solved. Chapter Three rejects the conclusions of the Rassmussen Report and certainly any suggestion that a melt-down cannot happen. At the time of writing no major reactor accident has occurred. There have been many close shaves. If just one reactor failed, the consequences, as described in Chapter Three, for many thousands of people, would be tragic. In a way the fate of the survivors, who would be at high risk for cancers and genetic damage, would be worse than that of the dead. Dale Bridenbaugh has repeatedly said that in his opinion a major accident is only a matter of time.

The Australian writer, Frank Hardy, reported a conversation he had had with an American nuclear engineer, travelling on a plane between Israel and Iran in December 1976. The engineer said that Iranian technicians were being trained in crash courses in reactor building and maintenance in Israel for the time when the Iranian nuclear industry got under way. He expressed the fear that, because technicians and tradesmen in Iran and other countries rapidly developing nuclear power, did not have the experience, skill or regulation to reach the level of competence required in reactor building, Iran's reactors would be unsafe. 'If one of their reactors blows they will blame us [the US]', he said.

Some argue that if an accident is to occur it had better be soon so that nuclear power can be stopped; these people do not usually live within range of a reactor. The last thing anti-nuclear people want to see is a disaster. They do not want their case proved right by the suffering of thousands of people, nor the decision to stop nuclear power made as a reaction to disaster, the nuclear-free world to be born in blood. Action must be taken before the predicted disaster occurs, especially to prevent the sale and construction of reactors to countries lacking the technical workforce to build and maintain them, or the political stability to use them responsibly, and the development of the most dangerous reactor of all, the fast breeder.

The unborn countless generations will be the inheritors of nuclear effluent produced by the present generation. Civilizations will rise and fall; languages will change and be lost; new political and social systems, as yet unthought of, will develop. And the geological process of change in the earth's crust will continue in ways that cannot be accurately predicted. Through all this the radioactive wastes will live on, guarded or unguarded, probably lost and forgotten, but still active, still lethal, the relic of a few decades of the twentieth century which will affect life for century upon century to come. No one can predict absolutely that a safe method of disposal will never be found; nor can the nuclear men predict that it will. For the sake of nuclear power, a leap in the dark, the earth could be poisoned for a million years and more.

At a uranium meeting in Sydney one participant suggested that if all the plutonium was sealed in bombs we would solve the problem of its disposal. Certainly nuclear war would end all argument. The proliferation of nuclear weapons is the lid on the coffin of the pro-nuclear case. No safeguards, no treaties, are ever likely to prevent nations and terrorists making and exploding bombs once plutonium from the fuel of hundreds of reactors is circulating around the globe. To sell uranium to the world in its present state is, says Greg Woods, like giving a continuous supply of first-class whisky to an alcoholic and then expecting him to sign the pledge of temperance. None of the experts on safeguards, like William Epstein and the men of the IAEA, can come up with any solution to the problem of enforcing them within sovereign states. People in Australia have been deluded into thinking that safeguards are a solid assurance that our uranium will not be made into bombs. That is an illusion. If nuclear power were completely clean, safe and cheap (as it patently is not) then the proliferation of weapons *alone* would be good reason to keep uranium in the ground. The danger of nuclear war is ever present and increasing with every country that newly acquires nuclear power. The concept of peaceful nuclear energy was a vain dream that is turning into a nightmare. The idea that the military and commercial atom can be split has been finally and firmly knocked on the head. Australia is now standing on the edge of a precipice deciding whether to jump. If it turns on its heel and marches purposefully in a new direction, it will not be alone. And its decision could illumine the history of the twentieth century, as a point of light in this dark century of greed for material and ideological dominance. Rarely has a country had such an opportunity for great example.

People have to decide whether to work for a nuclear world or a nuclear-free world. Both have been described. The first requires acceptance of an all-pervading fear of nuclear war breaking out at any time, all the health hazards associated with the industry and its wastes, expenditure of vast sums of money on building, maintaining and guarding nuclear plant, and increasingly centralized and secret government which will abrogate individual rights.

Instead the long process of extrication from the nuclear nightmare could begin. It would mean working towards a world in which energy would not be spent on goods that were thrown away to become a pollution problem in themselves. It would mean development of energy sources suitable to a humane way of life, which would guarantee individuals a great measure of independence, which would fulfil western civilization's ideal of individual freedom, the supposed basis of our law and democratic institutions. Sources such as solar and wind power can do this, as described in Chapter Six, both in developed and developing countries. By working on these alternatives, Australia could also make a real contribution to Third World needs.

So powerful is the case against nuclear power that one would expect, all things being equal, for it to be more widely accepted than it already is. Chapter Seven has described how the developed nations have ploughed deeper and deeper into the nuclear mire and what a world of difference there is between the values and tactics of pro-nuclear and anti-nuclear people.

Nuclear energy has been described by Amory Lovins as 'the technology of the future whose time is past'. It is the unrealized dream of the last world war's generation of nuclear technologists who are now either dead or old men. Because those at the head of government institutions and academic establishments are usually the middle-aged and old, the bases of their policies are frequently a couple of decades behind the current needs and knowledge of the time. In 1945 these men had a vision of unlimited cheap nuclear power. In their old age they cannot admit that this is now a miasma. Instead they train new generations of nuclear men in institutions which they control, to carry on the nuclear programme despite the doubts and problems that have emerged in the last twenty years. Public opinion polls in Australia over the last year have shown a significant trend in this respect. Those who support uranium export and nuclear power in the greatest numbers are over sixty years old. Those under twenty-five reject it by a large majority. It is, of course, of much greater immediate significance to a twenty-year-old than to a seventy-year-old whether nuclear power is safe enough to be the energy source for the future. The seventy-year-olds can speak of the year 2000, of the scenario in twenty, thirty or fifty years, with a degree of blithe detachment, for they are unlikely to have to live through it.

The deep concern felt over uranium export by young Australian people, together with their determination to find out the facts behind the nuclear propaganda and act upon them, is most encouraging. They may eventually lead us out of the nuclear trap, away from the tired visions of old men to a safer and freer future.

Another reason for the hold the nuclear men still have on the public mind is the realization, born of the Arab oil embargo, that the world cannot go on for much longer using up the world's finite fossil fuel reserves. The horrors of 'freezing in the dark', when the last cake-mixer and electric toothbrush are stilled, are described with gusto by the nuclear men, for the rocketing price of oil has saved their bacon, as far as propaganda is concerned. They pose as energy saviours of the world and denounce the opposing lobby for supporting the use of fossil fuels with all the pollution that that entails. It cannot be said too strongly that conservationists are as concerned about oil and petrol pollution as about nuclear hazards and have fought campaigns to enforce pollution controls and less polluting and energy-consuming methods of using fossil fuels. The nuclear men, on the other hand, though attacking the use of fossil fuels, are deeply involved in the oil industry. The oil companies, as pointed out in Chapter Seven, are moving into nuclear development in a big way. The Pan-continental mine at Jabiluka is partly financed by Getty Oil. The nuclear people cannot honestly pretend a moral aversion to the polluting oil industry when the finance and products of that industry, both in large quantities, are used to develop nuclear power.

Nuclear power is not the solution to the energy crisis. Unless fast breeders are developed, uranium itself would run out by the end of the century. But not one commercial fast breeder is working in the world at the moment, since they are beset by severe technical problems and are so potentially dangerous that on detection of a fault the whole plant must be closed down. It has been shown that nuclear power is an unreliable source of energy. It has become cripplingly expensive. Like the dinosaur, it is collapsing under its own weight.

Therefore it is a matter of the utmost urgency that a huge effort is made in research and development of energy conservation and supply from other sources – sources that cannot be centralized under the control of monopolies. The nuclear period of the 1950s and 1960s may one day be regarded by historians as an aberration, a grave mistake that will take many years to rectify. Economically and technically the nuclear industry is sick. It should be painlessly put out of its misery and full attention given to assisting the growth of that neglected but healthy and promising young infant, solar energy.

Tony Grey, the Chairman of Pancontinental, has said recently that the argument is not about nuclear power but about lifestyles. Sir Philip Baxter has said that those who oppose nuclear power would condemn mankind to returning to caves. Lang Hancock says mining is the basis of civilization. The nuclear power argument certainly is about lifestyle, about the social organization that would be determined by the politics and economics of a nuclear or non-nuclear society. The authors of this book reject a 'civilization' based on the rape of the world's resources for the benefit of a few in the present generation, but causing untold suffering to the many, long into the future. They do not define civilization, as does Lang Hancock, in terms of the amount of plastic, steel and electrical hardware with which many attempt, and fail, to make life more comfortable. Rather, they are concerned at the misery of the people of poor nations who do not get their fair share, or a share at all, of the world's wealth; and at the grey faces of those who work in concrete cities, choking their way backwards and forwards to work through the pollution of their own vehicles' exhaust pipes; and about the physical and mental strain this lifestyle imposes on people.

Their vision of the future is one of true cooperation between nations and people to spread the benefits of our earth to all, working together to solve the enormous problems of war, poverty and over-population, to channel our intelligence, energy and concern into this creative path and to do so by organizing our resources and social systems in such a way that every person is able to become more self-sufficient, independent and, therefore, free. This civilization does not come out of a mine. It follows from the release of human creative ability and an appreciation of man's dependence on, and intimate biological relationship with, all life on earth.

It is essential therefore to use every possible means to turn history in this direction. Pressure must be exerted by individuals on governments and officials. Each individual can take part in this. The greatest weapon in the battle is information. The lies of the nuclear men cannot prevail if they are exposed in the public forum. Again, each person can take part in the spread of information. The anti-uranium movement in Australia has grown from very small beginnings by these means – by researching the facts, by distributing information, by visiting politicians, making representation to the Fox Inquiry and government committees, writing letters to newspapers. This is happening all over the world. Every single Australian can, by doing these simple things, help to turn the tide against nuclear power. No one is alone and no action, however small, is irrelevant. Nothing in history has ever been achieved for the good of the people unless they themselves have demanded it, and with a loud voice. In Australia it is still possible to speak out. Many are doing so. In Australia and overseas the groundswell of protest is rising, a groundswell built up from many individual actions. If people demand something better for their children than a nuclear future they can and will get it. If they sit on the beach in silence, the hazards described in this book will be their legacy to the future.

There have been many creeds throughout human history, all promising happiness, freedom and the salvation of the world. None has yet succeeded. But at this turning point in human destiny people are realizing for the first time how delicate is the balance of nature, how close the human race is to upsetting the balance. The knowledge that the human species faces extinction at its own hands makes the petty and illogical quarrels between nations and races pale into insignificance. All over the world people are in revolt against such pettiness and there is surely a longing to build a world in which all can enjoy the beauty of the earth unshadowed by the mushroom cloud of fear. This is not idealism but realism. The choice is clear. It is either between a nuclear world in which the rich, industrialized nations continue on their present exploitative path, or a world based on the safety and survival of all species. Uranium is the test. Australia's choice on this issue can strongly affect, one way or the other, human survival.

This is our case. We present it to the people so that they, as is their inalienable right, may decide.

Afterword

International renunciation must start somewhere. Australia is in the key position to start it. The choice was intended to be given us by the compilers of the Fox Report. Will the Fraser government allow us any voice at all on this ultimate issue?

The implications of the warnings given in this first report are only beginning to sink into the minds of Australians. They are so alarming that it might seem incredible that any government could make a quick decision. The commissioners themselves have said that this is a decision for the ordinary man to make.

To commit even a part of our uranium deposits for export to a politically unstable world is a terrifying responsibility. It will affect the future of everyone in the world. Before this, all other environmental questions pale.

We have the power to decide to begin this move towards international renunciation. We have also the power to take action which may buy, for a few hundred million dollars, final disaster for the world.

Judith Wright
Extract from a speech of 11 November 1976 in reply to the Australian government's announcement on uranium export

Appendix A
First Fox Report

Principal Findings and Recommendations

1. The hazards of mining and milling uranium, if those activities are properly regulated and controlled, are not such as to justify a decision not to develop Australian uranium mines.
2. The hazards involved in the ordinary operations of nuclear power reactors, if those operations are properly regulated and controlled, are not such as to justify a decision not to mine and sell Australian uranium.
3. The nuclear power industry is unintentionally contributing to an increased risk of nuclear war. This is the most serious hazard associated with the industry. Complete evaluation of the extent of the risk and assessment of what course should be followed to reduce it involve matters of national security and international relations which are beyond the ambit of the Inquiry. We suggest that the questions involved are of such importance that they be resolved by Parliament. In Chapters 15 and 16 we have gone as far as the terms of reference and the evidence permit in examining the courses open and in making suggestions.
4. Any development of Australian uranium mines should be strictly regulated and controlled, for the purposes mentioned in Chapter 16.
5. Any decision about mining for uranium in the Northern Territory should be postponed until the Second Report of this Commission is presented.
6. A decision to mine and sell uranium should not be made unless the Commonwealth Government ensures that the Commonwealth can at any time, on the basis of considerations of the nature discussed in this Report, immediately terminate those activities, permanently, indefinitely or for a specified period.

7. Policy respecting Australian uranium exports, for the time being at least, should be based on a full recognition of the hazards, dangers and problems of and associated with the production of nuclear energy, and should therefore seek to limit or restrict expansion of that production.
8. No sales of Australian uranium should take place to any country not party to the NPT. Export should be subject to the fullest and most effective safeguards agreements, and be supported by fully adequate back-up agreements applying to the entire civil nuclear industry in the country supplied. Australia should work towards the adoption of this policy by other suppliers.
9. A permanent Uranium Advisory Council, to include adequate representation of the people, should be established immediately to advise the Government, but with a duty also to report at least annually to the Parliament, with regard to the export and use of Australian uranium, having in mind in particular the hazards, dangers and problems of and associated with the production of nuclear energy.
10. The Government should immediately explore what steps it can take to assist in reducing the hazards, dangers and problems of and associated with the production of nuclear energy.
11. Policy with regard to the export of uranium should be the subject of regular review.
12. A national energy policy should be developed and reviewed regularly.
13. Steps should be taken immediately to institute full and energetic programs of research and development into (a) liquid fuels to replace petroleum and (b) energy sources other than fossil fuels and nuclear fission.
14. A program of energy conservation should be instituted nationally.
15. The policy of the Government should take into account the importance to Australia, and the countries of the world, of the position of developing countries concerning energy needs and resources.

Our *final recommendation* takes account of what we understand to be the policy of the Act under which the Inquiry was instituted. It is simply that there should be ample time for public consideration of this Report, and for debate upon it. We therefore recommend that no decision be taken in relation to the foregoing matters until a reasonable time has elapsed and there has been an opportunity for the usual democratic processes to function, including, in this respect, parliamentary debate.

October 1976

Appendix B
Treaty on the Non-Proliferation of Nuclear Weapons

(Signed at London, Moscow, and Washington on 1 July 1968.)

The States concluding this Treaty, hereinafter referred to as the 'Parties to the Treaty',

Considering the devastation that would be visited upon all mankind by a nuclear war and the consequent need to make every effort to avert the danger of such a war and to take measures to safeguard the security of peoples,

Believing that the proliferation of nuclear weapons would seriously enhance the danger of nuclear war,

In conformity with resolutions of the United Nations General Assembly calling for the conclusion of an agreement on the prevention of wider dissemination of nuclear weapons,

Undertaking to co-operate in facilitating the application of International Atomic Energy Agency safeguards on peaceful nuclear activities,

Expressing their support for research, development and other efforts to further the application, within the framework of the International Atomic Energy Agency safeguards system, of the principle of safeguarding effectively the flow of source and special fissionable materials by use of instruments and other techniques at certain strategic points,

Affirming the principle that the benefits of peaceful applications of nuclear technology, including any technological by-products which may be derived by nuclear-weapon States from the development of nuclear explosive devices, should be available for peaceful purposes to all Parties to the Treaty, whether nuclear-weapon or non-nuclear-weapon States.

Convinced that, in furtherance of this principle, all Parties to the Treaty are entitled to participate in the fullest possible exchange of scientific information for, and to contribute alone or in co-operation with other States to, the further development of the applications of atomic energy for peaceful purposes

Declaring their intention to achieve at the earliest possible date the cessation of the nuclear arms race and to undertake effective measures in the direction of nuclear disarmament.

Urging the co-operation of all States in the attainment of this objective,

Recalling the determination expressed by the Parties to the 1963 Treaty banning nuclear weapon tests in the atmosphere, in outer space and under water in its Preamble to seek to achieve the discontinuance of all test explosions of nuclear weapons for all time and to continue negotiations to this end,

Desiring to further the easing of international tension and the strengthening of trust between States in order to facilitate the cessation of the manufacture of nuclear weapons, the liquidation of all their existing stockpiles, and the elimination from national arsenals of nuclear weapons and the means of their delivery pursuant to a treaty on general and complete disarmament under strict and effective international control,

Recalling that, in accordance with the Charter of the United Nations, States must refrain in their international relations from the threat or use of force against the territorial integrity or political independence of any State, or in any other manner inconsistent with the Purposes of the United Nations, and that the establishment and maintenance of international peace and security are to be promoted with the least diversion for armaments of the world's human and economic resources,

HAVE AGREED AS FOLLOWS:

Article I

Each nuclear-weapon State Party to the Treaty undertakes not to transfer to any recipient whatsoever nuclear weapons or other nuclear explosive devices or control over such weapons or explosive devices directly, or indirectly; and not in any way to assist, encourage, or induce any non-nuclear-weapon State to manufacture or otherwise acquire nuclear weapons or other nuclear explosive devices, or control over such weapons or explosive devices.

Article II

Each non-nuclear-weapon State Party to the Treaty undertakes not to receive the transfer from any transferor whatsoever of nuclear weapons or other nuclear explosive devices or of control over such weapons or explosive devices directly, or indirectly; not to manufacture or otherwise acquire nuclear weapons or other nuclear explosive devices; and not to seek or receive any assistance in the manufacture of nuclear weapons or other nuclear explosive devices.

Article III

1. Each non-nuclear-weapon State Party to the Treaty undertakes to accept safeguards, as set forth in an agreement to be negotiated and concluded with the International Atomic Energy Agency in accordance with the Statute of the International Atomic Energy Agency and the Agency's safeguards system, for the exclusive purpose of verification of the fulfilment of its obligations assumed under this Treaty with a view to preventing diversion of nuclear energy from peaceful uses to nuclear weapons or other nuclear explosive devices. Procedures for the safeguards required by this article shall be followed with respect to source or special fissionable material whether it is being produced, processed or used in any principal nuclear facility or is outside any such facility. The safeguards required by this article shall be applied on all source or special fissionable material in all peaceful nuclear activities within the territory of such State, under its jurisdiction, or carried out under its control anywhere.

2. Each State Party to the Treaty undertakes not to provide: (*a*) source or special fissionable material, or (*b*) equipment or material especially designed or prepared for the processing, use or production of special fissionable material, to any non-nuclear-weapon State for peaceful purposes, unless the source of special fissionable material shall be subject to the safeguards required by this article.

3. The safeguards required by this article shall be implemented in a manner designed to comply with article IV of this Treaty, and to avoid hampering the economic or technological development of the Parties or international co-operation in the field of peaceful nuclear activities, including the international exchange of nuclear material and equipment for the processing, use or production of nuclear material for peaceful purposes in accordance with the pro-

visions of this article and the principle of safeguarding set forth in the Preamble of the Treaty.

4. Non-nuclear-weapon States Party to the Treaty shall conclude agreements with the International Atomic Energy Agency to meet the requirements of this article either individually or together with other States in accordance with the Statute of the International Atomic Energy Agency. Negotiation of such agreements shall commence within 180 days from the original entry into force of this Treaty. For States depositing their instruments of ratification or accession after the 180-day period; negotiations of such agreements shall commence not later than the day of such deposit. Such agreements shall enter into force not later than eighteen months after the date of initiation of negotiations.

Article IV

1. Nothing in this Treaty shall be interpreted as affecting the inalienable right of all the parties to the Treaty to develop research, production and use of nuclear energy for peaceful purposes without discrimination and in conformity with articles I and II of this Treaty.

2. All the Parties to the Treaty undertake to facilitate, and have the right to participate in, the fullest possible exchange of equipment, materials and scientific and technological information for the peaceful uses of nuclear energy. Parties to the Treaty in a position to do so shall also co-operate in contributing alone or together with other States or international organizations to the further development of the applications of nuclear energy for peaceful purposes, especially in the territories of non-nuclear-weapon States Party to the Treaty, with due consideration for the needs of the developing areas of the world.

Article V

Each Party to the Treaty undertakes to take appropriate measures to ensure that, in accordance with this Treaty, under appropriate international observation and through appropriate international procedures, potential benefits from any peaceful applications of nuclear explosions will be made available to non-nuclear-weapon States Party to the Treaty on a non-discriminatory basis and that the charge to such Parties for the explosive devices used will be as low as possible and exclude any charge for research and development. Non-nuclear-weapon States Party to the Treaty shall be able to obtain such benefits, pursuant to a special international agreement or agreements, through an appropriate international body with adequate representation of non-nuclear-weapon States. Negotiations on this subject shall commence as soon as possible after the Treaty enters into force. Non-nuclear-weapon States Party to the Treaty so desiring may also obtain such benefits pursuant to bilateral agreements.

Article VI

Each of the Parties to the Treaty undertakes to pursue negotiations in good faith on effective measures relating to cessation of the nuclear arms race at an early date and to nuclear disarmament, and on a treaty on general and complete disarmament under strict and effective international control.

Article VII

Nothing in this Treaty affects the right of any group of States to conclude regional treaties in order to assure the total absence of nuclear weapons in their respective territories.

Article VIII

1. Any Party to the Treaty may propose amendments to this Treaty. The text of any proposed amendments shall be submitted to the Depositary Governments which shall circulate it to all Parties to the Treaty. Thereupon, if requested to do so by one third or more of the Parties to the Treaty, the Depositary Governments shall convene a conference, to which they shall invite all the parties to the Treaty, to consider such an amendment.

2. Any amendment to this Treaty must be approved by a majority of the votes of all the parties to the Treaty, including the votes of all nuclear-weapon States Party to the Treaty and all other Parties which, on the date the amendment is circulated, are members of the Board of Governors of the International Atomic Energy Agency. The amendment shall enter into force for each Party that deposits its instrument of ratification of the amendment upon the deposit of such instruments of ratification by a majority of all the Parties, including the instruments of ratification of all nuclear-weapon States Party to

the Treaty and all other Parties which, on the date the amendment is circulated, are members of the Board of Governors of the International Atomic Energy Agency. Thereafter, it shall enter into force for any other Party upon the deposit of its instrument of ratification of the amendment.

3. Five years after the entry into force of this Treaty, a conference of Parties to the Treaty shall be held in Geneva, Switzerland, in order to review the operation of this Treaty with a view to assuring that the purposes of the Preamble and the Provisions of the Treaty are being realized. At intervals of five years thereafter, a majority of the Parties to the Treaty may obtain, by submitting a proposal to this effect to the Depositary Governments, the convening of further conferences with the same objective of reviewing the operation of the Treaty.

Article IX

1. This Treaty shall be open to all States for signature. Any State which does not sign the Treaty before its entry into force in accordance with paragraph 3 of this article may accede to it at any time.

2. This Treaty shall be subject to ratification by signatory States. Instruments of ratification and instruments of accession shall be deposited with the Governments of the Union of Soviet Socialist Republics, the United Kingdom of Great Britain and Northern Ireland and the United States of America, which are hereby designated the Depositary Governments.

3. This Treaty shall enter into force after its ratification by the States, the Governments of which are designated Depositaries of the Treaty, and forty other States signatory to this Treaty and the deposit of their instruments of ratification. For the purposes of this Treaty, a nuclear-weapon State is one which has manufactured and exploded a nuclear weapon or other nuclear explosive device prior to 1 January 1967.

4. For States whose instruments of ratification or accession are deposited subsequent to the entry into force of this Treaty, it shall enter into force on the date of the deposit of their instruments of ratification or accession.

5. The Depositary Governments shall promptly inform all signatory and acceding States of the date of each signature, the date of

deposit of each instrument of ratification or of accession, the date of the entry into force of this Treaty, and the date of receipt of any requests for convening a conference or other notices.

6. This Treaty shall be registered by the Depositary Governments pursuant to Article 102 of the Charter of the United Nations.

Article X

1. Each Party shall in exercising its national sovereignty have the right to withdraw from the Treaty if it decides that extraordinary events, related to the subject-matter of this Treaty, have jeopardized the supreme interests of its country. It shall give notice of such withdrawal to all other Parties to the Treaty and to the United Nations Security Council three months in advance. Such notice shall include a statement of the extraordinary events it regards as having jeopardized its supreme interests.

2. Twenty-five years after the entry into force of the Treaty, a conference shall be convened to decide whether the Treaty shall continue in force indefinitely, or shall be extended for an additional fixed period or periods. This decision shall be taken by a majority of the Parties to the Treaty.

Article XI

This Treaty, the Chinese, English, French, Russian and Spanish texts of which are equally authentic, shall be deposited in the archives of the Depositary Governments. Duly certified copies of this Treaty shall be transmitted by the Depositary Governments to the Governments of the signatory and acceding States.

IN WITNESS WHEREOF the undersigned, duly authorized, have signed this Treaty.

DONE in triplicate at the cities of Washington, London and Moscow, this first day of July, one thousand nine hundred and sixty-eight.

References

Chapter Two

1 Back translation of Gunwinggu tape made by Jacob Mayinggul and Albert Balmana helped by Joseph Dumarda, James Balmana and Meryl Rowe, Oenpelli, 20 and 21 January 1976.

Chapter Three

2 J. P. Holdren (University of California, Berkeley), *Radioactive Pollution of the Environment by the Nuclear Fuel Cycle*, reproduced by Friends of the Earth, Australia, 1974.

Chapter Four

3 S. H. Smiley, 'Waste Management – Licensing and Criteria', in *Nuclear Technology*, December 1974.

4 R. W. Ramsey, G. H. Daley *et al.*, 'Overview of Management Programs for Pu-Contaminated Solid Waste in the USA', in OECD/NEA, *Management of Plutonium Contaminated Solid Wastes*, 1974, p. 65.

5 W. D. Rowe & W. F. Holcomb, USEPA, 'The Hidden Commitment of Nuclear Wastes' in *Nuclear Technology*, December 1974.

6 OECD/NEA, *Radioactive Waste Management in Western Europe*, Paris, 1971, p. 287.

7 OECD/NEA, *Radioactive Waste Management in Western Europe*, Paris, 1971, p. 73.

8 OECD/NEA, *Radioactive Waste Disposal Operation in the Atlantic*, 1967.

9 Rowe & Holcomb, USEPA, *op. cit.*

10 Ramsey *et al*, OECD/NEA, *op. cit.*

11 A. B. Lovins & J. H. Price, *Non-Nuclear Futures*, Ballinger, Cambridge, Mass., 1975, p. 69.

12 Battelle North West Laboratories, 'Advanced Waste Management Studies', BNWL-1900, 1974, p. 13.

13 W. D. Bishop & C. D. Hollister, 'Seabed Disposal – Where to Look', in *Nuclear Technology*, December 1974.
14 W. C. McClain & A. L. Boch, 'Disposal of Radioactive Waste in Bedded Salt Formations', in *Nuclear Technology*, December 1974.
15 USAEC Division of Waste Management and Transportation, 'High Level Radioactive Waste Management Alternatives', WASH-1297, May 1974.
16 Premier's Department, South Australia, 'Second Interim Report of the Uranium Enrichment Committee', 1976.

Chapter Five

17 M. Willrich & T. Taylor, 'Nuclear Theft: Risks and Safeguards', Ballinger (Lippincott), Cambridge, Mass., 1974.
18 Quoted in the *Congressional Record* – Senate 29 May 1974, S9203.
19 US Senator Proxmire, *Congressional Record* – Senate 25 June 1974, S11503.
20 R. W. Ayres, 'Policing Plutonium: the Civil Rights Fallout', *Harvard Civil Rights – Civil Liberties Law Review*, vol. 10, p. 411.

Further Reading

The literature on uranium and nuclear power is vast. Guidance through the maze can be found in Walter Patterson's bibliography to his book *Nuclear Power*. Here we simply mention some recent and readily available literature which has contributed substantially to the text of this book.

Birch, Charles, *Confronting the Future*, Penguin, Australia, 1976.
Epstein, William, *The Last Chance*, Macmillan/The Free Press, 1976.
Flood, M. and Grove-White, R., *Nuclear Prospects*, Friends of the Earth, in association with the Council for the Protection of Rural England and the National Council for Civil Liberties, London, 1976.
Hayes, D., Barrett, N. and Falk J., *Red Light for Yellowcake*, Friends of the Earth, Australia, 1977.
Holdren, J. P., *Radioactive Pollution of the Nuclear Fuel Cycle*, reproduced by Friends of the Earth, Australia, 1974.
Jungk, R., *Brighter than a Thousand Suns: A Personal History of the Atomic Scientists*, Penguin, Harmondsworth, 1976.
Lovins, A. B., and Price, J. H., *Non Nuclear Futures*, Ballinger, New York, London, 1975.
Lovins, A. B., *World Energy Strategies*, Friends of the Earth, New York and London, 1975.
Messel, H., and Butler, S. T., *Solar Energy*, Shakespeare Head, Sydney, 1974.
Novick, S., *The Careless Atom*, Delta, New York, 1970.
Patterson, Walter C., *Nuclear Power*, Penguin, Harmondsworth, 1976.
Ranger Uranium Environmental Inquiry First Report, AGPS, Canberra, 1976. *Second Report*, AGPS, Canberra, 1977.

Royal Commission on Environmental Pollution, Sixth Report, *Nuclear Power and the Environment*, HMSO, London, 1976.

Schumacher, E. F., *Small is Beautiful*, Blond and Briggs, London, 1973.

Willrich, M. and Taylor T., *Nuclear Theft: Risks and Safeguards*, Ballinger (Lippincott), Cambridge, Mass., 1974.

Index

Aborigines, 43, 48
   land rights, 51, 52, 60, 210
   at Ranger Inquiry, 53-62
   sacred sites, 28, 29, 50, 53
Alligator Rivers Region, 27, 36, 38, 41, 42, 50
ALP policy, 4, 199
Anthony, Doug, 197
Atomic Energy (Special Constables) Act, 157
Australian Atomic Energy Commission (AAEC), 3, 26, 47, 48, 156
   and Jervis Bay, 189, 191, 193
   as a political force, 162, 189, 190
   at the Ranger Inquiry, 131, 141, 191
   and Rum Jungle, 27, 31-6, 48

Baxter, Sir Philip, 127, 141, 189, 193, 197
Bjelke-Petersen, J., 193
Bridenbaugh, Dale, xi, 211

Conzinc Rio Tinto, 26, 198, 201

ecological effects of uranium mining, 42, 50, 97
Einstein, Albert, 131, 132
energy conservation, 177-9
energy strategy, 160, 175-6
enrichment process, 18, 19, 99
Environment Protection (Impact of Proposals) Act, 1
Environmental Protection Agency (US), 18
Epstein, William, 129, 132, 134, 138, 211

Flowers Report, 112, 120, 124, 147, 162, 172, 175-6
fuel fabrication, 19, 99, 192

genetic effects of radiation, 8, 13, 123, 210
Gorton, John, 198

Hancock, Lang, 194
health hazards, 7-17
Hiroshima, 13, 14, 15, 23, 25, 131

IAEA safeguards, 130, 140-44
International Commission on Radiological Protection, 16, 18, 35

Jabiluka, 27, 28, 40, 201, 214
Jervis Bay, 65, 189

Kakadu National Park, 40

Material Unaccounted For (MUF) Factor, 141, 149
Mary Kathleen, 3, 26, 27, 201
mutations, 15, 123, 210

Nabarlek, 27
Newman, Kevin, 131
nuclear fuel cycle, 8, 12, 66
nuclear industry
   cost, 169-71, 202
   insurance, 192-3
   involvement of oil companies in, 192, 202, 214
   leasing, 205
   multi-national control, 197-9
   origin in Manhattan Project, 187, 190
   price fixing, 199
   propaganda, 188, 189, 194, 210
Nuclear Non-Proliferation Treaty (NPT), 131-4, 138
   aims, 133
   new members, 138, 139
   non-members, 133, 134, 140
   weaknesses, 133, 135, 136, 138
NPT Review Conference, 130, 144
nuclear testing, French, 134
nuclear war, 127, 129

Pancontinental Mining Ltd, 28, 54, 201, 214
Patterson, Walter, xi, 3, 65, 95
plutonium, 5, 99, 101, 108, 130, 162
   toxicity, 9, 10, 19, 22-3, 191
   in weapons, 101, 129-30

policing the plutonium economy, 153
protest abroad, 127, 206

Queensland Mines, 201

radiation, 16, 17
   background, 99
   as cause of cancer, 10, 11, 14
radon, 99, 103, 104
   effect on miners, 10, 16
Ranger Company, 5, 46, 47, 52
Ranger Mine, 27, 42, 49
Ranger Uranium Environmental Inquiry, 1-4, 52, 55
   First Report, 3, 22, 30, 43, 117, 124, 135, 136, 163
   estimate of uranium reserves, 30, 165
   NPT and IAEA safeguards, 132, 138, 140-43
   terms of reference 2, 161, 195
   on terrorism, 146-50, 152-3
   Second Report, 2, 3, 17, 40, 48
reactors, 19, 68, 99, 190
   accident history, 89-95
   accident predictions, 84, 87, 88
   accident potential, 82-4, 87, 89, 211-12
   breeder reactors, 80, 81, 99, 124, 168-9, 193, 205, 214
   decommissioning, 102, 109
   how they work, 67, 69-73, 77-79, 81, 82, 84
   sabotage of, 89, 90
   types of, 65, 69, 70-76
reprocessing, 21-2, 98, 99, 100, 101

Rum Jungle, 26, 27, 31-3, 56
Russell, Bertrand, 132

safeguards, 17, 102, 130, 132, 138, 140, 141, 143, 153, 212
solar energy, 159
    advantages of, 159, 171, 181-3
    cost, 171, 173, 175, 184
    lack of research, 163
    its many forms, 165, 166
    problems, 174
    and social change, 186, 206
    as solution to energy problems, 180
stockpiling, 142

tailings, 17, 18, 32, 45-6, 47, 99, 102
terrorism, 145, 149-50
terrorist bomb construction, 145
Titterton, Sir Ernest, 193, 196

uranium
    Australian reserves, 1, 27, 28, 30, 196
    history of development, 25, 31, 41
    mining procedures, 45
    and the Third World, 195, 206

uranium miners
    health hazards to, 98
Uranium Producers' Forum, 3, 193

wastes, 20, 102
    burial, 106, 107
    disposal in ice caps, 97, 117
    disposal in salt, 120-22
    disposal in seabed, 97, 117, 119-20
    disposal in space, 117
    high level, 98, 101, 108, 110, 115, 116, 124
    leaks, 98, 108, 109, 110
    length of life, 99, 101, 108, 110, 115-16, 118
    low level, 98, 99, 101, 106
    from mining, 98
    ocean dumping, 104, 105
    reactor effluents, 108
    solidification, 113
    transmutation, 118
    vitrification, 114
weapons industry, 26, 107, 127

Yeelirrie, 29
yellowcake, 17, 44, 135